The Incompleteness of Deductive Logic

And Its Consequences for Epistemic Engineering

John-Michael Kuczynski

First edition.
Published by Zhi Systems.
ISBN: 978-1-968752-05-7
Library of Congress Control Number: 2026936054

Table of Contents

Part III: Natural Language and Metalinguistic Hierarchies

Part IV: Breaking the Stranglehold of Recursion

Part IV: Breaking the Stranglehold of Recursion (Continued) Scientific Discovery and Non-Recursive Rationality

Part V: Rationality as Pattern Recognition

Part V: Rationality as Pattern Recognition (Continued)

Part 0: Introduction

This book proves a mathematical theorem and traces its philosophical consequences across multiple domains. The theorem is technical, but its implications are revolutionary. It shows that rationality cannot be reduced to recursive rule-following, and that recognizing this fact dissolves centuries of pseudo-problems in philosophy, vindicates modern artificial intelligence, and opens up vast new research territories.

The Core Theorem

Main Result: The class of all recursive, truth-preserving logics is not recursively enumerable.

What this means: There is no algorithm that can list all possible formal logical systems. Recursion cannot recursively enumerate itself. Formalism cannot formalize its own boundaries.

This is not just another incompleteness result. Gödel showed that no single formal system can capture all arithmetic truths. This theorem shows something deeper: no recursive procedure can even survey the space of formal systems themselves. The incompleteness goes all the way up.

Why This Matters

The theorem has immediate consequences for:

- **Logic and Mathematics**: "Formal truth" is not a property statements have intrinsically, but a relation to particular logical systems—and the space of these systems is provably non-enumerable.

- **Philosophy of Language**: Natural languages are hierarchies ($E_0 \cup E_1 \cup E_2 \cup \ldots$) where each level can refer to lower levels but not to itself. Syntax can be recursive, but semantics necessarily includes non-recursive elements like ostensive reference, naming, and metalinguistic awareness.

- **Artificial Intelligence**: Neural networks succeed precisely because they're NOT recursive systems. The false equation "Intelligence = Recursive Computation" doomed Good Old-Fashioned AI (GOFAI). Modern AI works because it can approximate both recursive and non-recursive rational capacities.

- **Epistemology**: Hume's problem of induction dissolves. Induction doesn't need justification by deduction because rationality is broader than recursivity. Pattern recognition is basic, self-justifying, and constitutive of rational thought.

- **Philosophy of Science**: Scientific discovery is rational but non-recursive. Popper was wrong to conclude that because discovery isn't algorithmic, it must be irrational. Newton's inference to universal gravitation, Darwin's recognition of natural selection—these are paradigmatic examples of non-recursive rationality.

The False Equation

For over a century, multiple domains have been infected by a poisonous assumption:

RATIONAL = RECURSIVE

This equation says: to make sense of something is to recursively formalize it. To understand X is to give recursive rules for X. Reasoning is following algorithms. Knowledge is what can be recursively justified.

This assumption came from:

- Logic's prestige (Frege, Russell)

- Computability theory (Turing, Church)

- Chomsky's revolution (over-generalizing syntactic recursion)

- Logical positivism (meaning = verification procedure)

And it caused catastrophic failures:

- Hume's induction problem (induction seems irrational because it's not deductive)

- Popper's irrationalism (discovery can't be rational—no algorithm for it)

- Wittgenstein's paradox (rule-following leads to infinite regress)

- GOFAI's collapse (intelligence requires symbolic rules)

- Compositional semantics dogma (meaning must be recursive)

- Moral reductionism (ethics needs universal principles)

This book shows the equation is false.

Rationality is broader than recursivity. There are two modes of rational thought:

1. **Recursive rationality**: Following explicit rules, deductive inference, compositional operations, algorithmic procedures. Important but LIMITED.

2. **Non-recursive rationality**: Pattern recognition, abductive inference, semantic grounding, concept formation, insight, perception, judgment. Equally important, BROADER domain.

Both are rational. Neither reduces to the other. And the main theorem proves that rationality cannot recursively characterize its own scope—just as formalism cannot formalize its own boundaries and language cannot linguistically characterize itself.

The Positive Program

Once we reject Rational = Recursive, we can make genuine progress by recognizing and studying non-recursive rationality:

In Philosophy of Science: Study how scientists actually reason. Formalize abduction, unification, pattern recognition. Recognize that discovery is rational (just not algorithmic).

In Epistemology: Develop non-foundationalist theories. Understand perceptual justification as direct pattern recognition. Explore particularist and coherentist approaches.

In Philosophy of Mind: Investigate non-computational mental capacities. Study concept formation, insight, creativity. Connect to neuroscience of pattern recognition.

In Philosophy of Language: Recognize that linguistic competence = UG (recursive syntax) + X (non-recursive semantic capacities). Accept that compositional semantics cannot capture ostension, naming, and metalinguistic awareness.

In Ethics: Develop particularist moral theories. Study moral perception and practical wisdom. Stop searching for recursive moral principles that don't exist.

In Mathematics: Vindicate mathematical intuition. Study non-formal aspects of mathematical knowledge. Recognize that understanding proofs requires seeing why steps follow, not just checking symbols.

In Artificial Intelligence: Embrace neural networks and statistical approaches. Stop trying to force everything into recursive molds. Study how to approximate non-recursive rationality.

The Structure of This Book

Part I: The Incompleteness of Logic proves the main theorem through diagonalization and injective embedding, showing that the class of recursive logics is not recursively enumerable.

Part II: Summary and Extensions reformulates the result multiple ways, extends it to language (showing the class of all languages is not r.e.), and situates it within the hierarchy of incompleteness results from Cantor through Gödel.

Part III: Natural Language and Metalinguistic Hierarchies applies the theorem to linguistics, showing that natural languages are stratified hierarchies where syntax can outrun semantics, the metalinguistic series is not recursive, and language cannot linguistically characterize itself.

Part IV: Breaking the Stranglehold of Recursion establishes the syntax/semantics distinction (syntax is recursive, semantics is not), shows why computational linguistics cannot fully model semantics, vindicates neural networks, refutes the false equation Rational = Recursive, and liberates multiple domains from recursive tyranny.

Part V: Rationality as Pattern Recognition redefines rationality as the skilled capacity to recognize truth-relevant patterns, shows how this dissolves Hume's problem, demonstrates that pattern recognition recovers all deductive functions, refutes Wittgenstein's rule-following paradox, and establishes the positive program across all affected domains.

What You Need to Know

The book moves from technical proof to philosophical application. Part I is mathematical and requires familiarity with recursion theory and formal logic. Parts II-V are progressively more philosophical and

accessible, though they build on the technical foundation.

Key prerequisites:

- Basic recursion theory (recursive functions, recursive enumerability)

- Elementary logic (formal systems, proof theory)

- Willingness to question deeply held assumptions about rationality

What you don't need:

- Advanced mathematics beyond basic set theory and computability

- Prior agreement with any particular philosophical school

- Patience for academic hedging and qualification

This book makes strong claims and defends them directly. The arguments are rigorous but not baroque. The style is clear and uncompromising. If you want philosophically transformative work that's also mathematically sound, read on.

The Central Insight

The tradition mistook a special case (recursive/deductive reasoning) for the whole of rationality. This mistake created pseudo-problems that have plagued philosophy for centuries. The main theorem proves that this mistake was inevitable given the false equation Rational = Recursive, because:

Rationality cannot recursively characterize its own scope.

Just as:

- Formalism cannot formalize its own boundaries

- Recursion cannot recursively enumerate itself

- Language cannot linguistically characterize language So

too:

- Rationality transcends recursive specification

This is not a limitation but a liberation. Once we recognize the full breadth of rational capacity—including the non-recursive capacities that actually make science, language, thought, and intelligence work—we can stop forcing everything into recursive molds and make genuine progress.

Rationality is seeing, not calculating.

That's what this book proves, and that's what changes everything.

Part I: The Incompleteness of Logic

A Recursion-Theoretic Generalization of Gödel's Theorem

Abstract

This chapter proves that the class of all deductive logics—understood as recursively defined, truth-preserving sets of statements—is not recursively enumerable. By generalizing Gödel's incompleteness result, we show that the very space of logical systems cannot be exhaustively captured by any recursive procedure. The argument proceeds in two steps: first, by proving via diagonalization that the set of total recursive functions is not recursively enumerable; and second, by constructing an injective embedding of total recursive functions into distinct truth-preserving logics.

1. Definitions and Framework

We work within the framework of classical first-order arithmetic with standard model $\mathbb{N}$ as our background theory of truth.

Definition 1.1 (Logic): A logic is a pair (S, Φ), where:

- S is a finite set of base statements (axioms)

- Φ is a total recursive function from strings to strings

The logic-extension L generated by (S, Φ) is the smallest set K such that:

1. $S \subseteq K$

2. For every x in K, $\Phi(x) \in K$

Definition 1.2 (Truth-Preserving): A function Φ is truth-preserving if whenever x is a true statement, $\Phi(x)$ is also a true statement (under the intended interpretation).

Definition 1.3 (Recursive Logic): A logic (S, Φ) is recursive if Φ is a total recursive function and S is finite.

This framework captures the essence of formal systems: they start with axioms and generate theorems through recursive operations. What we call "logic" here includes propositional logic, modal logic, type theory, set theory, and any other recursively axiomatized system. The definition is deliberately minimal to maximize generality.

2. The Non-Enumerability of Total Recursive Functions

Theorem 2.1: The set of all total recursive functions is not recursively enumerable.

Proof: Suppose, toward contradiction, that there exists a recursive enumeration $\{\Phi_0, \Phi_1, \Phi_2, ...\}$ of all total recursive functions from $\mathbb{N}$ to $\mathbb{N}$.

Define a new function F: $\mathbb{N} \to \mathbb{N}$ by:

$F(n) = \Phi_n(n) + 1$

Since each Φ_n is total and recursive, and addition is recursive, F is itself a total recursive function. However, F differs from every function in the enumeration: for each n, we have:

$F(n) = \Phi_n(n) + 1 \neq \Phi_n(n)$

Therefore $F \neq \Phi_n$ for all n.

Thus F is a total recursive function not in our enumeration, contradicting the assumption that the enumeration was complete.

Therefore, the set of total recursive functions is not recursively enumerable. ∎

Remarks: This is a pure diagonalization argument in the style of Cantor and Turing. The key insight is that totality prevents enumeration—if we tried to enumerate total recursive functions, we could diagonalize out of the enumeration while remaining total. This is the engine that drives the entire book's argument.

3. Injective Embedding of Recursions into Logics

Not every recursive function directly generates statements. However, we can construct an injective mapping from total recursive functions to truth-preserving logics that preserves distinctness.

Construction 3.1 (Canonical Logic Encoding): For each total recursive function f: $\mathbb{N} \to \mathbb{N}$, we construct a corresponding logic $L\_f = (S\_f, \Phi\_f)$ as follows:

Base axioms S_f:

- "$f(0) = k_0$" where $k_0 = f(0)$

Recursive operator Φ_f:

- Input: "$f(n) = k_n$"

- Output: "$f(n) = k_n \land f(n+1) = k_{n+1}$" where $k_{n+1} = f(n+1)$

The logic-extension generated by L_f contains all statements of the form:

- "$f(0) = k_0$"

- "$f(0) = k_0 \land f(1) = k_1$"

- "$f(0) = k_0 \land f(1) = k_1 \land f(2) = k_2$"

- etc.

Each statement encodes the complete computational history of f up to some point n.

Lemma 3.2: The mapping $f \mapsto L\_f$ is injective.

Proof: Suppose $f \neq g$. Then there exists some n_0 such that $f(n_0) \neq g(n_0)$.

The logic L_f contains the statement "$f(0) = f(0) \land ... \land f(n_0) = f(n_0)$", which encodes the value $f(n_0)$. The logic L_g contains the statement "$g(0) = g(0) \land ... \land g(n_0) = g(n_0)$", which encodes the value $g(n_0)$. Since $f(n_0) \neq g(n_0)$, these statements are distinct. Therefore, the sets L_f and L_g differ, so $L\_f \neq L\_g$. ∎

Lemma 3.3: Each logic L_f is truth-preserving under the standard interpretation of arithmetic.

Proof: Each statement in L_f is a conjunction of arithmetic equations of the form "$f(i) = k_i$" where k_i is the actual value computed by f at i. Since these equations state true facts about f's behavior, and conjunction preserves truth, every statement in L_f is true. ∎

Remarks: This construction is the crucial bridge. We're not claiming that these logics are interesting or useful—only that they exist, are recursive, preserve truth, and are distinguishable. The encoding is mechanical but serves its purpose: translating the non-enumerability of functions into the non-enumerability of logics.

4. Main Result

Theorem 4.1: The class of all recursive, truth-preserving logics is not recursively enumerable.

Proof: By Construction 3.1 and Lemma 3.2, there exists an injective function from the set of total recursive functions to the set of recursive truth-preserving logics.

By Theorem 2.1, the set of total recursive functions is not recursively enumerable.

Since there is an injection from a non-r.e. set into the set of recursive logics, the set of recursive logics cannot be recursively enumerable (if it were, we could pull back the enumeration to enumerate total recursive functions).

∎

Corollary 4.2: There is no recursive definition of "recursive definition."

Corollary 4.3: There is no recursive procedure that generates all recursive procedures.

Corollary 4.4: Recursivity cannot formally capture its own extent.

These corollaries follow immediately: if we could recursively enumerate recursive procedures, we could enumerate recursive logics (since logics are generated by recursive procedures). But Theorem 4.1 shows this is impossible.

5. Formal Truth as a Relational Concept

Traditionally, philosophers have sought to define "formal truth" as an intrinsic syntactic property—something about the logical form of a statement that makes it true independent of interpretation.

This paper suggests a different conception:

Definition 5.1 (Formal Truth - Relational): A statement S is formally true if and only if there exists a truth-preserving recursive logic L such that $S \in L$.

Formally:

FormallyTrue(S) $\Leftrightarrow$ ∃L [L is truth-preserving $\land$ L is recursive $\land$ S $\in$ L]

This reframes formal truth not as an intrinsic property but as a relationship between a statement and a truth-preserving recursive system.

Corollary 5.2: The predicate "formally true" is not recursively enumerable.

Proof: If "formally true" were r.e., we could enumerate all formally true statements. But each formally true statement belongs to some recursive logic, and different logics can be distinguished by their member statements. This would allow us to enumerate the recursive logics themselves, contradicting Theorem 4.1. ∎

Remarks: This relational conception has profound implications. There is no algorithm that can list all formal truths. No procedure can generate or recognize all formal systems. The space of formal systems transcends formal enumeration. Formalism cannot formalize its own boundaries.

6. Philosophical Consequences

6.1 The Historical Confusion About Logical Form

Historically, certain expressions ("and," "or," "not," "all," "some") have been classified as "logical constants" while others ("knows," "believes," "possibly") have not. This distinction has often been presented as tracking something deep about meaning or form.

This paper suggests a deflationary view: expressions count as "logical" when we have identified truth-preserving recursive rules governing them. The apparent special status of classical logical constants reflects merely the fact that we discovered workable recursive systems for them early (via truth tables, natural deduction, etc.).

There is no intrinsic syntactic mark of "logicality"—only the pragmatic fact that some expressions admit readily discoverable recursive axiomatizations.

6.2 The Limit of Formal Knowledge

Gödel's original incompleteness theorem showed that no single recursive axiom system can capture all

arithmetic truths. This result generalizes that insight: no recursive meta-procedure can generate all recursive logical systems.

Just as Gödel placed a limit on what any particular formal system can prove, this result places a limit on what the space of formal systems itself can be: it cannot be recursively surveyed.

The hierarchy of limitation:

- **Cantor**: Some sets are larger than others (diagonal out of enumeration)

- **Gödel**: Some truths escape any single formal system (diagonal sentence)

- **This theorem**: The space of formal systems itself escapes enumeration (diagonal function)

Each level shows a deeper impossibility.

6.3 Dissolution of the Formality Problem

The question "What makes a truth formally true?" has been a central problem in philosophy of logic. This paper suggests the question was misconceived.

Formal truth is not a monadic property ($\checkmark$ or X) but a relation to a recursive proof system. Since the space of such systems is not recursively enumerable, there is no decision procedure for formality itself.

This dissolves rather than solves the traditional problem: there is no unified essence of "formal truth" to be discovered, only an infinite, non-enumerable plurality of recursive truth-preserving systems.

The key insight: Philosophers sought the form of formality. But formality has no form. It cannot recursively characterize itself.

7. Conclusion

We have proven:

1. The set of total recursive functions is not recursively enumerable (Theorem 2.1)

2. Total recursive functions can be injectively embedded into truth-preserving recursive logics (Construction 3.1, Lemma 3.2)

3. Therefore, the class of recursive logics is not recursively enumerable (Theorem 4.1)

4. Consequently, "formal truth" (understood relationally) is not a recursively enumerable predicate (Corollary 5.2)

This establishes a recursion-theoretic boundary to formal knowledge that is more fundamental than Gödel's original result. While Gödel showed that no single logic is complete, we have shown that the space of logics itself defies algorithmic completion.

The dream of a unified, recursive characterization of logical truth—a master algorithm that could generate or recognize all formal systems—is provably unrealizable.

Formalism cannot formalize its own boundaries.

This is not a practical limitation requiring better techniques. It is a principled impossibility proven by diagonalization. And as we will see in subsequent chapters, this impossibility has cascading consequences across philosophy, linguistics, artificial intelligence, and our understanding of rationality itself.

References

Church, A. (1936). An unsolvable problem of elementary number theory. *American Journal of Mathematics*, 58(2), 345–363.

Gödel, K. (1931). Über formal unentscheidbare Sätze der Principia Mathematica und verwandter Systeme I.

Monatshefte für Mathematik und Physik, 38(1), 173–198.

Kleene, S. C. (1952). *Introduction to Metamathematics*. North-Holland.

Rogers, H. (1967). *Theory of Recursive Functions and Effective Computability*. McGraw-Hill.

Turing, A. M. (1936). On computable numbers, with an application to the Entscheidungsproblem. *Proceedings of the London Mathematical Society*, 2(42), 230–265.

Part II: Summary and Extensions

Core Theorem

Main Result: The class of all recursive, truth-preserving logics is not recursively enumerable.

Proof Structure

The argument proceeds in three steps:

Step 1 - Diagonalization (Theorem 2.1): The set of all total recursive functions is not recursively enumerable. Proof via diagonalization: construct $F(n) = \Phi_n(n) + 1$, which differs from every Φ_n in any purported enumeration.

Step 2 - Canonical Encoding (Construction 3.1): For each total recursive function $f: \mathbb{N} \to \mathbb{N}$, construct a logic $L\_f = (S\_f, \Phi\_f)$:

- Base axiom: "$f(0) = k_0$" where $k_0 = f(0)$

- Recursive operator Φ_f: Input "$f(n) = k_n$" → Output "$f(n) = k_n \wedge f(n+1) = k_{n+1}$"

- Generates: "$f(0) = k_0$", "$f(0) = k_0 \wedge f(1) = k_1$", "$f(0) = k_0 \wedge f(1) = k_1 \wedge f(2) = k_2$", etc.

Lemma 3.2: The mapping $f \mapsto L\_f$ is injective (different functions → different logics)

Lemma 3.3: Each L_f is truth-preserving

Step 3 - Main Result (Theorem 4.1): Therefore, the class of recursive logics is not r.e. (injection from non-r.e. set)

The beauty of this proof is its simplicity. We don't need elaborate machinery. Just diagonalization and an encoding function. The impossibility follows directly.

Key Reframings

Multiple Ways to State the Main Result

The theorem admits several equivalent formulations, each illuminating a different aspect:

1. **There is no recursive definition of "recursive definition"**

   - Any attempt to formally characterize what counts as a formal characterization fails

2. **There is no recursive procedure that generates all recursive procedures**

   - The space of algorithms transcends algorithmic enumeration

3. **Recursivity cannot formally capture its own extent**

   - Recursive methods have a blind spot: themselves

4. **The concept of "algorithm" cannot be algorithmically delimited**

   - We cannot mechanically determine what counts as mechanical

5. **Recursion cannot recursively enumerate itself**

   - The class of recursive functions outruns any recursive enumeration of that class

6. **There is no formal characterization of formal truth**

   - Formality cannot formalize what formality is

These are not merely different phrasings. They're different angles on the same fundamental limitation.

The Core Insight

Formalism cannot formalize its own boundaries.

Any attempt to define "what counts as formal" will either:

- **Miss some formal systems** (be incomplete), or
- **Use non-formal (semantic, meta-level) resources** (be circular)

This is not a practical difficulty requiring better definitions. It's a principled impossibility proven by diagonal argument. The boundary of formalism cannot itself be formalized.

Proven Theorems (Solid)

From the main result, several rigorous corollaries follow:

1. **The set of all total recursive functions is not r.e.** (Theorem 2.1)

   - Proven via diagonalization

2. **The class of all recursive, truth-preserving logics (S, Φ) is not r.e.** (Theorem 4.1)

   - Follows by injection from #1

3. **The class of all consistent extensions of PA is not r.e.**

   - Inject functions as axiom sets encoding computational histories

4. **The class of all sound proof systems is not r.e.**

   - Inject via computational histories encoded as inference rules

Each theorem shows the same pattern: wherever there's enough structure to encode recursive functions, non-enumerability appears.

Extension to Language

Definition: A language is (approximately) a recursive expression-class, or more simply, a recursive statement-class.

Theorem: The class of all languages is not recursively enumerable.

Proof: If languages = recursive statement-classes, then the class of all languages = the class of all recursive statement-generating systems. By our main theorem, this class is not r.e. ∎

Consequence: There is no recursive definition of "language" itself.
This has immediate implications:

Implications

Linguistics: Universal grammar cannot be algorithmically characterized. If UG is supposed to capture what all possible human languages share, and languages are recursive systems, then UG cannot recursively specify its own domain.

Programming language theory: Cannot enumerate all possible programming languages. No meta-language can generate all languages. Language design requires non-formal insight.

Formal semantics: No master theory of meaning. Any recursive semantic theory will miss possible languages. Semantics outstrips formalization.

Philosophy of language: Cannot linguistically characterize language. Any characterization of "what language is" must step outside language itself. The limit of language cannot be stated in language.

Comparison with Classical Results

The Hierarchy of Incompleteness

Three major results form a hierarchy of increasing depth:

Cantor ($\mathbb{R} > \mathbb{N}$):

- **What it shows**: The reals cannot be enumerated by the naturals

- **Method**: Diagonalization

- **Domain**: Set theory, cardinality

- **Insight**: Some infinities are bigger than others

Gödel (Incompleteness):

- **What it shows**: No consistent formal system can prove all arithmetical truths

- **Method**: Self-reference + arithmetization

- **Domain**: Formal systems, arithmetic

- **Insight**: Formal systems can't capture arithmetic truth

This Theorem (Logics not r.e.):

- **What it shows**: No recursive procedure can enumerate all formal systems

- **Method**: Injection from total recursive functions

- **Domain**: The space of formal systems itself

- **Insight**: The space of formal systems transcends formal enumeration

Each result reveals a deeper level of incompleteness.

The Deepening

Gödel: "We can't complete arithmetic"

- Formal systems can't capture arithmetic truth

- You need intuition/insight to see unprovable truths

- Some arithmetic statements require non-formal recognition

This Theorem: "We can't even survey the space of possible completions"

- Formal systems can't even capture the space of formal systems

- You need intuition/insight to recognize what counts as a valid formal system

- Meta-formal judgment requires non-formal resources

The incompleteness goes up a level. It's not just that we can't formalize everything within arithmetic. We can't even formalize what "formalization" encompasses.

The Regress

The hierarchy reveals a regress of incompleteness:

1. **Arithmetic needs intuition** (Gödel)

   - To see truths that no system can prove

2. **Recognizing valid systems needs intuition** (This theorem)

   - To identify what counts as a formal system

3. **Recognizing when intuition is needed needs intuition** (This theorem, meta-level)

   - To see that formalization requires non-formal judgment

Punchline: Incompleteness isn't just a feature of arithmetic. It's a feature of formalism itself.

No formal framework can self-certify. No recursive procedure can enumerate all recursive procedures. Recursion can't close over itself.

At every level, formality requires something beyond formality to ground it.

Practical Applications

Cantor's Diagonal ($\mathbb{R} > \mathbb{N}$)

- **Computer science**: Undecidability of halting problem

- **Cryptography**: Random number generation, infinite key spaces

- **Information theory**: Channel capacity limits

- **Formal verification**: Proving impossibility results

Gödel's Incompleteness

- **AI safety**: Formal verification has principled limits

- **Automated theorem proving**: Recognizing unprovable statements

- **Foundations of mathematics**: Program correctness has limits

- **Cybersecurity**: Proving absence of vulnerabilities is impossible

This Theorem (Logics not r.e.)

- **Meta-verification**: Can't enumerate all verification frameworks

- **AI alignment**: Can't algorithmically survey all possible reasoning systems

- **Formal methods**: Tool selection requires non-formal judgment

- **Programming language design**: Can't mechanically generate all type systems

- **Standards/certification**: No algorithmic meta-standard for formal standards

- **Linguistics**: Universal grammar can't be recursively characterized

- **Automated reasoning**: No master algorithm for recognizing valid inference systems

- **Knowledge representation**: Can't enumerate all possible knowledge frameworks

The practical import: At the meta-level, formalization requires judgment. No algorithm can tell you which formal method to use. Tool selection is irreducibly non-formal.

Why This Matters

Gödel vs This Theorem

Gödel's result is about limits within systems.

- Any particular formal system has blind spots

- But maybe we could enumerate all systems?

This result is about limits of the space of systems itself.

- No—we can't even enumerate all systems

- The meta-level has the same structure as the object level

- Incompleteness goes all the way up

This is deeper than Gödel. It shows incompleteness is not a bug that could be fixed by going meta. It's a structural feature of formalism that appears at every level.

Five Key Implications

1. Formalism can't bootstrap itself

Any attempt to formally characterize "what counts as formal" fails. Either:

- The characterization misses some formal systems (incomplete)

- The characterization uses non-formal resources (circular)

Formality requires meta-formal judgment.

2. No master logic

There's no algorithm that generates or recognizes all logical systems. The dream of a unified formal framework

—one logic to rule them all—is dead.

The space of logical systems is irreducibly plural.

3. Recursive inadequacy goes all the way up

It's not just that arithmetic is incomplete. Incompleteness infects the metalevel. You can't escape by going meta. The regress is infinite.

Meta-formalization requires non-formal resources just as much as object-level formalization does.

4. Philosophy of logic consequences

"Logical truth" can't be recursively defined. It's genuinely plural, irreducibly so. There is no essence of "the logical" to be captured.

Logic is plural. Formality is relational. Truth-preservation comes in non-enumerable varieties.

5. AI/computability limits

No computer can enumerate all possible reasoning systems. Creative formal thinking is provably non-algorithmic.

Building new formal systems requires non-recursive insight.

The Philosophical Upshot

Gödel: Proof needs something beyond proof (semantic truth/intuition).

- To see that the Gödel sentence is true, you need semantic intuition
- Proof can't self-certify

This Theorem: Formality needs something beyond formality (meta-formal judgment).

- To recognize what counts as a valid formal system, you need non-formal judgment
- Formalism can't self-certify Incompleteness

at a higher level of abstraction. The pattern:

- Gödel showed that formal systems need something outside themselves (semantic truth)
- This theorem shows that formalism itself needs something outside itself (meta-formal judgment)
- At each level, closure is impossible

The space of rational thought transcends recursive specification.

This sets up the rest of the book. If rationality cannot be exhaustively formalized, what is rationality? If formalism requires non-formal resources, what are those resources? The answer: pattern recognition. Non-recursive rationality. And that's what the remaining parts develop.

Transition to Part III

We've proven the theorem and traced its immediate implications. Now we extend the result to natural language, showing that languages are hierarchies ($E_0 \cup E_1 \cup E_2 \cup \ldots$) where syntax can be recursive but semantics necessarily includes non-recursive elements. This will set up Part IV's argument that the false equation Rational

= Recursive has infected multiple domains, and Part V's positive program showing that rationality is fundamentally pattern recognition—broader than recursivity, deeper than deduction, and capable of what formalism provably cannot achieve.

Part III: Natural Language and Metalinguistic Hierarchies

The Problem of Self-Reference in Language

The Foundational Circularity

Consider E = the class of all English expressions.

The expression "E contains every English expression" appears to be a grammatical English sentence. But there's a problem:

- For "E" to refer, E must already exist as a completed totality

- But E is defined as the class of all English expressions

- So E's existence depends on sentences containing "E"

- But sentences containing "E" depend on E already existing

Conclusion: We cannot have E ∈ E. No language can contain statements that presuppose the language's own existence as a completed totality.

This parallels the Foundation Axiom in set theory: sets cannot presuppose their own existence. Just as $S \notin S$ for sets, a language cannot make its own totality an object of discourse within itself.

The problem is not merely technical. It reveals something fundamental about the structure of natural language.

Natural Language as Hierarchy

The Solution: Stratification

Natural languages are not a single recursive expression-class. They are **hierarchies** of such classes, stratified by self-reference levels.

English = $E_0 \cup E_1 \cup E_2 \cup ... \cup E_n \cup ...$

Where:

E_0 = Base recursive expression-class (doesn't mention "English" or reference the language itself)

- "Snow is white"

- "$2 + 2 = 4$"

- "Recursive functions are computable"

E_1 = First metalevel (mentions E_0 as an object)

- "Every expression of E_0 is in E_0"

- "E_0 is recursive"

- "E_0 contains 'snow is white'"

E_2 = Second metalevel (mentions E_1)

- "E_1 can describe E_0"

- "Every expression of E_1 is in E_1"

$E_3, E_4, ...$ and so on indefinitely

How the Hierarchy Works

At each level E_n:

- You can make statements about E_0, E_1, ..., E_{n-1}
- You cannot make well-grounded statements about E_n itself (that requires E_{n+1})
- No vicious circularity because the referent always exists at a lower level

This is Tarski's hierarchy for truth, but applied to language-as-object rather than just truth predicates.

Syntax vs. Semantics in the Hierarchy

A Critical Distinction

A language = syntax + semantics

An expression truly belongs to a language L only if:

1. It's **syntactically well-formed** in L (built from L-vocabulary using L-rules) AND

2. It has **semantic content** in L (refers, has truth-conditions, means something)

The Puzzling Case

Consider the phrase: **"the totality of English sentences"**

In E_0:

- ✓ Syntactically well-formed (built from E_0 words: "the", "totality", "of", "English", "sentences")
- ✓ Combined using E_0 grammatical rules (definite description)
- ✗ Has no semantics (E_0 doesn't exist as a completed object yet, so the phrase has no referent)

In E_1:

- ✓ Syntactically well-formed (inherited from E_0)
- ✓ Has semantics (E_0 now exists as a completed object that E_1 can refer to)

The Upshot

Syntax can run ahead of semantics.

You can build syntactically valid strings that await semantic grounding at a higher level.

- E_0 contains the **string** "the totality of English sentences"
- E_1 contains the **meaningful expression** "the totality of English sentences"

The string is proto-linguistic—it has grammatical form but no content until the metalevel provides the referent.

The hierarchy isn't just adding new vocabulary. It's **semantically activating** syntax that was already present but inert.

Generation of Metalevels

How E_{n+1} Arises from E_n

The transition from E_n to E_{n+1} involves:

1. Identifying syntactically well-formed but semantically empty strings in E_n

2. Treating E_n as a completed object

3. Adding metalinguistic apparatus to refer to E_n

4. Semantically activating previously inert syntax

This is **not a recursive operation** internal to E_n. It requires:

- Stepping outside the system

- Introducing new terms ("E_0", "E_1", etc.) as neologisms

- Meta-formal judgment about what constitutes reference to the totality

The Critical Question: Is $(E_0, E_1, E_2, ...)$ Recursive?

For the series to be recursive, we'd need a recursive function F such that:

- $F(0) = E_0$

- $F(n+1) = E_{n+1}$

But generating E_{n+1} from E_n requires:

- Semantic judgment (which strings get semantically activated?)

- Metalinguistic creativity (introducing new referring terms)

- Recognition of what constitutes "talking about E_n as a totality"

None of these operations appear algorithmically specifiable.

Consequence: The series $(E_0, E_1, E_2, ...)$ is not recursive.

This is a direct application of the main theorem: recursivity cannot recursively enumerate itself. Here, we cannot recursively enumerate the metalinguistic levels because each level-jump requires non-recursive semantic operations.

The Infinite Regress Problem

The Meta-Metalanguage

Suppose (arguendo) that the series $(E_0, E_1, E_2, ...)$ were recursive. What language are we using to say:

- "The series $(E_0, E_1, E_2, ...)$ is recursive"

- "Each E_{n+1} is generated from E_n by semantic activation"

- "The union $\bigcup_i E_i$ = English"

Call this metalanguage **E***.

But E* contains:

- All the E_i as objects of discourse

- Claims about the entire series

- Claims about the generating procedure

- Claims about recursivity itself

E* is not in any E_i

So either:

1. There's an external metalanguage E^* that's more powerful than any E_i

   - E^* generates E^{**} to talk about itself

   - E^{***} to talk about E^{**}

   - **Infinite regress...**

2. Or alternatively: E^* is not recursive because it would need to enumerate all the E_i, violating the main theorem.

The Fundamental Point

Even if we grant that $(E_0, E_1, E_2, ...)$ is recursive, the language in which we describe that recursivity cannot itself be part of the series.

Any attempt to give a complete account of the language hierarchy requires stepping outside that hierarchy.

This is exactly what the main theorem predicts: recursivity cannot recursively characterize itself. Similarly, a natural language hierarchy cannot recursively characterize its own structure.

Connection to Logicism

Why Logicism Failed

The logicist project (Frege, Russell) aimed to reduce all mathematics to logic. But they never had a clear conception of what "logic" meant.

They oscillated between incompatible views:

Logic = Topic-Neutral Generality (maximally formal/abstract truths)

- **Problem**: Cantor's theorem feels "logical" but makes mathematics incomplete

Logic = Recursive Closure of Tautologies (start with $\{P \lor \neg P, ...\}$, apply modus ponens)

- **Problem**: This is at most recursively enumerable, can't capture $\mathbb{R}$ or even full arithmetic

Logic = Whatever System I'm Using (second-order logic with comprehension)

- **Problem**: This is set theory in disguise, not logic

They wanted "analytically derivable from logical truths" but never clarified what "derivable" meant:

- Recursively provable? (No—Gödel)

- Semantically valid? (Not recursive)

- Intuitively evident? (Psychologism!)

The fundamental error: Conceptual incoherence about what 'logic' meant, not just technical failure.

Once Cantor proved $\mathbb{R}$ is uncountable (1891), logicism should have been dead if "logic" meant "recursively axiomatizable formal system."

Everything after was either:

- Calling set theory "logic" (terminological sleight-of-hand)

- Restricting to arithmetic and pretending that's what they meant

- Using second-order logic (which isn't recursively complete)

This theorem shows: They couldn't even recursively enumerate the candidate formal systems, so even the restricted project can't be carried out algorithmically.

Summary of New Results

Main New Theorems

1. **No language can enumerate all languages** (corollary of main theorem)

2. **Natural languages are hierarchies, not single recursive expression-classes**

   - English = $E_0 \cup E_1 \cup E_2 \cup \ldots$

   - Each level E_n allows reference to lower levels but not to itself

3. **Syntax can outrun semantics**

   - Expressions can be syntactically well-formed in E_n but semantically empty

   - They acquire meaning only at E_{n+1} when E_n becomes an object

4. **The metalinguistic series is not recursive**

   - Generating E_{n+1} from E_n requires non-algorithmic operations

   - Semantic judgment, metalinguistic creativity, recognition of totality-reference

5. **Metalinguistic ascent generates infinite regress**

   - Any language E* that describes the hierarchy stands outside it

   - This E* requires E**, which requires E***, etc.

   - Or: E* is not recursive (can't enumerate all E_i)

Philosophical Implications

No language can fully characterize itself

- Parallel to: No formal system can fully characterize formal systems

- Parallel to: Recursivity cannot recursively define itself

Natural language transcends recursivity

- Locally recursive (each E_n is recursive)

- Globally non-recursive (the union and the generating series are not)

Semantic grounding requires stepping outside

- To give meaning to self-referential expressions, you need a metalevel

- The metalevel itself requires a meta-metalevel

- No final level of complete semantic closure

The limits of language cannot be stated in language

- Wittgenstein's ladder, but mathematically proven

- Any statement of language's limits requires transcending those limits

Formalism cannot formalize its own boundaries

- This applies not just to logical systems but to language itself

- Natural language is the ultimate formal system that cannot self-certify

Technical Notes

What "Formal" Means

"Formal" $\neq$ "Arithmetical"

- **Arithmetical**: Statements about $\mathbb{N}$ (natural numbers), addition, multiplication

- **Formal**: Any recursively defined syntactic system (includes propositional logic, modal logic, set theory, type theory, etc.)

"Formal" $\supset$ "Arithmetical"

The theorem applies to ANY recursive, truth-preserving logic, not just arithmetic. This makes it MORE general than Gödel's result.

The Definition of Logic Used

Definition 1.1 (Logic): A logic is a pair (S, Φ), where:

- S is a finite set of base statements (axioms)

- Φ is a total recursive function from strings to strings

The logic-extension L generated by (S, Φ) is the smallest set K such that:

1. $S \subseteq K$

2. For every x in K, $\Phi(x) \in K$

Definition 1.2 (Truth-Preserving): A function Φ is truth-preserving if whenever x is a true statement, $\Phi(x)$ is also a true statement (under the intended interpretation).

Definition 1.3 (Recursive Logic): A logic (S, Φ) is recursive if Φ is a total recursive function and S is finite.

Why This Definition Works

This is a MINIMAL definition of a formal system:

- Starts with axioms (S)

- Has a recursive generation rule (Φ)

- Preserves truth

It's not the traditional definition of "logic" (with inference rules, connectives, etc.), but it captures the essence:

recursively defined, truth-preserving statement-generation.

The injection works precisely because this definition is minimal enough to encode arbitrary recursive functions.

Open Questions and Future Directions

Potentially Provable Extensions

Modal Logic:

- The class of all recursively axiomatized normal modal logics is not r.e.
- Requires: explicit injection construction + proof that modal structure is preserved

Epistemic Logic:

- The class of all recursively defined epistemic closure operators is not r.e.
- Requires: similar construction + verification

Type Systems:

- The class of all sound type systems is not r.e.
- Requires: encoding recursive functions into type judgments

Philosophical Extensions

The K Paradox (Epistemic Closure):

- Let K = set of all knowable truths
- Consider S: "I know all members of K"
- Can $S \in K$? No. Either contradiction or incompleteness.
- **Conclusion**: No maximal set of knowable truths
- **Connection**: Similar structure to main theorem, but different proof method

Critique of Neo-Logicism:

- Crispin Wright's neo-logicism uses second-order logic
- Second-order logic is not recursively axiomatizable
- **Therefore**: Not "logic" in the recursive sense
- Wright reduces arithmetic to something MORE POWERFUL (set theory in disguise)
- Not explanatory progress—circular

Critique of Epistemic/Deontic Logic:

- Standard axioms are empirically false:
    - $Kp \rightarrow KKp$ (positive introspection): false
    - $O(p) \rightarrow \Diamond(p)$ (ought implies can): false with counterexamples
- Most "applied" uses are window dressing on information-theoretic results

Areas Requiring Careful Treatment

Fuzzy Logic:

- Actually useful (engineering, control systems)
- Not really about "degrees of truth" but about modeling continuous properties
- Legitimate applications survive the critique

Mereology:

- "Parts of parts are parts" fails for functional parthood

- Transitivity only holds for spatial containment

- Most mereological "theorems" formalize intuitions without explaining them

Conclusion

We've shown that natural languages cannot be single recursive expression-classes. They must be hierarchies where:

1. Each level can refer to lower levels but not to itself

2. Syntax can be recursive at each level, but semantics requires non-recursive operations

3. The generating series itself is not recursive

4. Any attempt to characterize the hierarchy requires stepping outside it

This extends the main theorem from logic to language, showing that the same structural limitation appears wherever we try to make systems self-reflective.

The pattern is clear: Formalism cannot formalize its own boundaries. Recursion cannot enumerate itself. Language cannot linguistically characterize itself.

And as we'll see in Part IV, this pattern has infected multiple domains through the false equation Rational = Recursive, creating pseudo-problems that dissolve once we recognize non-recursive rationality.

Transition to Part IV

We've proven that:

- The space of logics is not recursively enumerable (Part I)

- This shows formalism cannot formalize its boundaries (Part II)

- Natural languages are hierarchies with non-recursive semantic operations (Part III)

Now we turn to the broader implications. If syntax is recursive but semantics is not, what does this mean for computational linguistics? If language requires non-recursive capacities, what does this mean for AI? And most importantly: if rationality includes irreducibly non-recursive elements, how do we understand rationality itself?

Part IV breaks the stranglehold of recursion by showing that the equation Rational = Recursive is false and has caused catastrophic failures across multiple domains.

Part IV: Breaking the Stranglehold of Recursion

Introduction

Parts I-III established that:

- The class of recursive logics is not recursively enumerable

- Formalism cannot formalize its own boundaries

- Natural languages are hierarchies with non-recursive semantic operations

Now we trace the implications across multiple domains. The central claim: a false equation has infected philosophy, linguistics, cognitive science, and AI for over a century:

RATIONAL = RECURSIVE

This equation says: to make sense of something is to recursively formalize it. To understand X is to give recursive rules for X. Reasoning is following algorithms. Knowledge is what can be recursively justified.

This equation is false. And recognizing its falsity liberates vast territories of inquiry.

The Syntax/Semantics Distinction

The Core Insight

Syntax is necessarily recursive (definitionally so). Semantics is not recursive.

This is the fundamental architecture of natural language.

Syntax: The Recursive Part

Syntax involves:

- Combining expressions according to recursive rules: "John" + "runs" $\rightarrow$ "John runs"

- Embedding: "Mary thinks [John runs]"

- Infinite generativity from finite means

This is what Chomsky's Universal Grammar (UG) captures:

- Merge operation

- Phrase structure rules

- Recursive composition

Syntax is compositional at the syntactic level:

- Well-formed structures built from well-formed parts

- Recursive generation

UG exists. Chomsky was right about this.

Semantics: The Non-Recursive Part

But meaning is not always compositional. You can use recursive syntax to build expressions whose semantics is non-recursive.

Key example: "The totality of expressions of English"
Syntactically:

- $\checkmark$ Built from E_0 vocabulary: "the", "totality", "of", "expressions", "of", "English"

- $\checkmark$ Combined by E_0 syntactic rules (definite description)

- $\checkmark$ Recursively generated

Semantically:

- ✗ Has no referent in E_0 (E_0 doesn't exist as a completed object yet)

- $\checkmark$ Only acquires meaning in E_1 (when E_0 becomes an object)

- ✗ Meaning is not a compositional function of part-meanings

- ✗ Non-recursive

Why Compositionality Fails

Standard semantic compositionality principle:

$$\llbracket \alpha\ \beta \rrbracket = f(\llbracket \alpha \rrbracket, \llbracket \beta \rrbracket)$$

The meaning of a complex expression is a function of:

- The meanings of its parts
- How they're syntactically combined

But "the totality of expressions of English" violates this:

- ✓ Each word has meaning in E_0
- ✓ Syntactic combination is well-formed
- ✗ But the whole phrase has no referent in E_0

This is not a bug. It's a feature of how natural language handles self-reference and metalinguistic expressions.

Other Examples of Semantic Non-Compositionality

"The present King of France" (Russell)

- Syntactically well-formed
- No referent (France has no king)
- Not a compositional failure—a reference failure

"The largest prime number"

- Syntactically well-formed
- No referent (no largest prime exists)
- Each word meaningful, syntax fine, but whole fails to refer

"This sentence is false" (Liar paradox)

- Syntactically well-formed
- Semantically unstable (no determinate truth value)
- Compositionality breaks down

But "the totality of expressions of English" is more fundamental:

- Not a mere reference failure (like "King of France")
- Not a logical impossibility (like "largest prime")
- A **level-mismatch**: The expression presuposes its own semantic grounding

The General Principle

For any natural language L with syntax S: You
can use S to construct expressions whose:

- Syntax is recursive (built by S's rules)

- Semantics is non-recursive (meaning/reference not compositionally determined)

This is essential to natural language, not a bug or edge case.

Why Formal Semantics Struggles

Formal semantics (Montague, Davidson, etc.) tries to give recursive semantic rules:

- ⟦John runs⟧ = runs(john)

- ⟦every man runs⟧ = $\forall x[man(x) \rightarrow runs(x)]$

This works for fragments of language where:

- All primitives have fixed referents

- Compositionality holds

- No metalinguistic expressions

- No naming/ostension

- No self-reference

But it fails for:

- Ostensive reference ("that child there")

- Naming acts ("Let 'Smith' refer to that child")

- Concept formation (creating new primitives like "quark")

- Metalinguistic statements ("English has 26 letters")

- Self-referential expressions

These are semantically essential but non-recursive.

Linguistic Competence: UG + X

Vindicating and Extending Chomsky

Chomsky discovered that language has a recursive core: Universal Grammar (UG). But UG is not the whole story.

Linguistic Competence = UG (recursive) + X (non-recursive)

What UG Provides (The Recursive Part)

UG gives you:

- **Syntax:** Combine "John" + "runs" → "John runs"

- **Recursion:** Embed clauses: "Mary thinks [John runs]"

- **Compositionality:** Meaning of syntactic wholes from parts

- **Infinite generativity:** From finite vocabulary, generate infinite sentences

This is the recursive computational system.

Chomsky was absolutely right about this. UG exists, it's innate, it's universal, it's computational.

What X Provides (The Non-Recursive Part)

But UG doesn't give you:

- **Ostensive/demonstrative reference**: "that child there"

- **Naming capacity**: "Let 'Smith' refer to that child"

- **Concept formation**: Creating new semantic primitives ("quark", "Brexit", "meme")

- **Metalinguistic awareness**: Talking about the language itself ($E_0 \to E_1$ hierarchy)

- **Semantic grounding**: Connecting language to world via perception, convention, social practice

These require **X**—the non-recursive supplement to UG.

What X Consists Of

The non-recursive component (X) includes:

Demonstrative/Ostensive Reference

- "That child there" $\to$ grounds reference perceptually

- Requires attentional/perceptual mechanisms

- Not recursive (not compositional)

- Not derivable from syntactic rules

Naming Capacity

- Creating new primitive referring terms

- "Let 'Smith' refer to that child"

- Socially coordinated (conventional)

- Not algorithmic, not rule-based

Concept Formation

- Creating new semantic primitives

- Not recursive combination of old concepts

- Genuine conceptual innovation (Newton's "force," Darwin's "natural selection")

Metalinguistic Awareness

- Talking about language itself

- Generating hierarchy ($E_0 \to E_1 \to E_2...$)

- Recognizing when expressions are semantically grounded vs. empty

Semantic Judgment

- Recognizing when expressions are meaningful vs. semantically empty

- Determining what new expressions should mean

- Meta-semantic capacity, not reducible to compositional rules

The Empirical Observation

The meta-semantic capacity is empirically obvious: Every

time someone:

- Names their newborn child
- Coins a new term for a discovered phenomenon
- Creates a neologism for a new concept
- Points and says "that thing there"

They are exercising X: the non-recursive meta-semantic capacity. This capacity is:

- **Universal** (all speakers have it)
- **Systematic** (not random or error-prone)
- **Essential** (without it, language couldn't adapt or refer to new entities)

Therefore: **X is part of linguistic competence, not mere performance.**

Competence vs. Performance Reconsidered

Chomsky distinguished:

- **Competence** = Internalized system (UG)
- **Performance** = Actual use (includes errors, context, pragmatics)

He relegated naming, ostension, pragmatics to performance.

But this is wrong.

The meta-semantic capacity (X) is:

- Systematic (not error-prone like performance)
- Universal (all speakers have it)
- Essential (constitutive of linguistic ability)

Therefore: **X belongs to competence, alongside UG.**

Why X Is Non-Recursive

The naming/ostension capacity is irreducibly non-recursive because:

Depends on demonstrative reference:

- "That child there" requires perceptual grounding
- Perception is not a linguistic/recursive operation

Requires social coordination:

- Getting others to accept "Smith" as name for child
- Social practices aren't algorithmic

Involves contingent facts:

- Who gets born, what gets discovered
- Not specifiable by fixed rules in advance

Creates new semantic primitives:

- Not recursive combination of existing meanings

- Genuine semantic expansion

By the main theorem: The class of recursive expression-classes is not r.e. Applied here: The meta-semantic capacity generates new recursive expression-classes (adding new primitives to vocabulary), but this generating process itself is not recursive.

The Two-Part Architecture

Natural language = Recursive core (UG) + Non-recursive supplement (X)

Both are:

- Essential to linguistic competence

- Universal across speakers

- Innate (in some sense)

But they differ in kind:

- **UG is computational/algorithmic**

- **X is meta-computational/non-algorithmic**

Why This Matters

This resolves longstanding puzzles:

Why formal semantics struggles:

- It tries to recursively specify something (full semantics) that includes non-recursive elements (X)

- It can model fragments (where X isn't operative) but not the whole

Why language feels both systematic and creative:

- Systematic: UG provides recursive structure

- Creative: X allows genuine semantic innovation

Why language can adapt to new situations:

- UG alone would be fossilized (fixed vocabulary, fixed meanings)

- X allows language to grow (new names, new concepts, new referents)

Why Chomsky was both right and incomplete:

- Right: There's a recursive computational core (UG)

- Incomplete: That's not the whole story (missing X)

Implications for Computational Linguistics

What Is Computational Linguistics?

Computational Linguistics (CL) has two branches:

Theoretical/Scientific CL:

- Goal: Understand language by building computational models

- Projects: Formalizing syntax, modeling semantics, simulating acquisition
- Question: Can language be computationally modeled?

Applied/Engineering CL (NLP):

- Goal: Build systems that process/generate language
- Applications: Translation, speech recognition, text generation, etc.
- Question: What can computers do with language?

Impact on Theoretical CL

The traditional assumption: "If we can computationally model syntax and semantics separately, we can model language."

The result of our analysis: This assumption is false because:

- Syntax is recursive (modelable by UG-based systems) ✓
- Semantics is not fully recursive (includes X) ✗
- Therefore, no purely computational model captures full linguistic competence ✗

Specific implications:

For syntax modeling:

- CL can successfully model UG (parsing, generation)
- This works because syntax is recursive
- No problem here ✓

For semantics modeling:

- CL tries to give recursive semantic rules (compositional semantics)
- But semantics requires:
    - Ostensive grounding ("that child there")
    - Naming capacity (adding new primitives)
    - Metalinguistic awareness ($E_0 \rightarrow E_1 \rightarrow E_2...$)
    - Concept formation (creating new semantic atoms)
- None of these are recursive
- Therefore, no complete computational semantics possible ✗

This isn't a practical limitation (need more data, better algorithms). It's a principled impossibility—by the main theorem, recursivity cannot capture its own extent, and semantic grounding transcends recursive specification.

Impact on Applied CL/NLP

The pragmatic question: Does NLP care about recursive vs. non-recursive semantics?

Answer: It depends.

What current NLP can do:

Modern NLP (especially LLMs like GPT, Claude) can:

- Generate syntactically well-formed text ✓

- Approximate semantic patterns from data ✓

- Translate, summarize, answer questions ✓

Why this works:

- They don't need true semantic grounding

- They pattern-match on massive corpora

- They approximate understanding via statistical correlation

What current NLP cannot do:

True ostensive reference:

- Can't point to "that child over there" and learn "Smith" refers to it

- Can't ground demonstratives in perception

- Can't connect language to world via demonstrative acts

- **Why not?** No perceptual grounding, no embodied reference

- **Note:** This limitation is about embodiment, not recursion

Genuine naming capacity:

- Can't genuinely decide "Let's call this new thing X"

- Can't create new primitive semantic rules

- Can't truly expand semantic base with genuinely new referents

- Models can coin new strings (e.g., "let's call it a 'flurb'"), but they can't:

  - Ground those strings in new referents

  - Truly add to their semantic foundation

- **Note:** This limitation might also be about embodiment/intentionality, not recursion

Metalinguistic awareness:

- Can't genuinely recognize own language as an object

- Can't generate the hierarchy $E_0 \rightarrow E_1 \rightarrow E_2$... with genuine understanding

- Can't understand "the totality of my expressions" as semantically grounded

- Models can simulate metalinguistic talk (e.g., "As an AI, I..."), but this is:

  - Pattern-matching on training data

  - Not genuine metalinguistic awareness

- **Note:** Whether this is a fundamental limitation is unclear

CRITICAL POINT

The limitations above are not because of the recursion/non-recursion distinction.

AI/Neural networks are not recursive systems (they're based on matrix operations, gradient descent, statistical learning—not recursive functions).

Therefore: **The main theorem (about recursive systems) does not directly constrain AI.**

Apples and bowling balls.

The question of whether AI can achieve:

- True semantic grounding

- Genuine naming capacity

- Authentic metalinguistic awareness

...is a different question about:

- Embodiment (connecting to world causally)

- Intentionality (whether statistical patterns constitute genuine understanding)

- Consciousness (whether there's "something it's like" to process information)

These are separate issues from recursion theory.

Neural Networks Vindicated

Our Work is FAVORABLE to AI, Not Threatening

The old (wrong) view:

- "Language is fundamentally recursive (Chomsky)."

- "Therefore, to model language, we need recursive/symbolic systems."

This led to:

- Classical AI (symbolic, rule-based, recursive)

- Good Old-Fashioned AI (GOFAI)

- Expert systems, logic-based reasoning

What our work shows:

- "Language includes recursive parts (syntax/UG) but also non-recursive parts (semantics/X)."

- Therefore: The recursive paradigm was over-applied

- We thought recursion was essential everywhere, but it's not

Implication for AI

Old worry: "Neural networks don't follow explicit recursive rules, so they can't really handle language."

Our response: "Good! Because language competence isn't purely recursive anyway. The meta-semantic capacity (X) was never recursive to begin with."

So: **Neural networks not being recursive is a feature, not a bug.**

They might actually be better suited to approximate the non-recursive aspects of language than rule-based systems ever were.

The Slogan Explained

"If neural networks were recursive, they couldn't do jackshit."

Meaning: If they were limited to recursive operations, they'd be stuck with GOFAI's limitations—unable to:

- Learn from unstructured data

- Recognize novel patterns

- Approximate non-recursive rational capacities

- Handle the non-compositional aspects of semantics

"Because they are non-recursive, there is nothing they cannot do."

Meaning: Being non-recursive frees them from the principled limitations on recursive systems. They can:

- Approximate any pattern (given enough data, capacity, training)

- Handle both recursive and non-recursive aspects of cognition

- Potentially achieve anything that doesn't require embodiment, genuine intentionality, or other non-computational prerequisites

- **No principled limitation from recursion theory**

This is why modern AI (neural/statistical) succeeded where classical AI (symbolic/recursive) failed.

What CL Should Recognize

Syntax ≠ Semantics:

- Syntax: Recursive, fully modelable

- Semantics: Includes non-recursive elements, not fully modelable by recursive rules

Competence Includes Non-Computational Elements:

- UG (recursive) is computable

- X (meta-semantic, non-recursive) may not be recursively specifiable

- But X might be approximable by non-recursive learning systems (neural networks)

Complete Recursive Semantics Is Impossible:

- Not a practical limitation (need more data, better algorithms)

- A principled impossibility (by main theorem and nature of semantic grounding)

Neural/Statistical Approaches Are Vindicated:

- Not inferior to symbolic/recursive approaches

- Actually better suited for non-recursive aspects of language

- The future of CL, not a temporary expedient

The False Equation: Rational = Recursive

The Poisonous Assumption

The equation that infected multiple domains:

"Rational = Recursive"

Or equivalently:

- "To make sense of X = To recursify X"

- "To understand X = To formalize X recursively"

- "Reasoning = Following recursive rules"

- "Knowledge = What can be recursively justified"

- "Intelligence = Recursive computation"

This equation is FALSE.

And recognizing its falsity liberates vast territories of inquiry.

Where the Equation Came From

Logic's prestige (Frege, Russell, early analytic philosophy):

- Logic is recursive (proof theory, axioms + inference rules)

- Logic seemed like the paradigm of rationality

- Therefore: rationality = recursivity (seemed to follow)

Computability theory (Turing, Church):

- Effective procedures = recursive functions (Church-Turing thesis)

- "Computable" became synonymous with "mechanical" and "rigorous"

- Non-recursive = incomputable = mysterious/irrational (seemed to follow)

Chomsky's revolution (Universal Grammar):

- Language has recursive structure (syntax)

- Language = paradigm of human cognitive capacity

- Generalized to: all cognitive competence is recursive

- **Over-extension of a local truth**

Verificationism (Logical Positivism):

- Meaning = verification procedure

- Verification = following recursive rules

- Therefore: meaningful = recursively specifiable

- Non-recursive = meaningless

Pattern: Taking a local truth (logic is recursive, syntax is recursive, some reasoning follows rules) and wildly generalizing (therefore ALL rationality is recursive).

The Consequences

This false equation infected multiple domains:

- **Philosophy of Science** → Popper's irrationalism

- **Epistemology** → Foundationalist anxiety

- **Philosophy of Mind** → Computationalism

- **Philosophy of Language** → Compositional semantics dogma

- **Ethics** → Moral reductionism

- **Mathematics** → Formalism vs. intuitionism wars

- **Artificial Intelligence** → GOFAI's failure

We'll examine each in detail in the next sections.

Transition to Remaining Sections

We've established:

1. Syntax is recursive, semantics is not (UG + X architecture)

2. Computational linguistics can model syntax but not full semantics (principled impossibility)

3. Neural networks are vindicated because they're non-recursive (feature, not bug)

4. The false equation Rational = Recursive has infected multiple domains

The remaining sections of Part IV will show how this false equation created catastrophic failures in:

- Scientific discovery (Popper's irrationalism)

- Neural networks and AI (vindication of non-recursive approaches)

- Multiple philosophical domains (liberation from recursive tyranny)

- The positive program (recognizing non-recursive rationality)

Part V will then develop the positive account: rationality is fundamentally pattern recognition, which recovers all deductive functions, dissolves Hume's problem, refutes Wittgenstein's paradox, and shows that seeing is more fundamental than calculating.

Part IV: Breaking the Stranglehold of Recursion (Continued) Scientific Discovery and Non-Recursive Rationality

Karl Popper on Scientific Discovery

Popper's Argument:

Premise 1: There is no recursive/algorithmic method for going from observational data to scientific theories.

Premise 2 (Implicit): Rationality = recursivity (following rules, logical inference)

Conclusion: Therefore, scientific discovery is irrational. There is no "logic of discovery," only a "logic of justification."

- **Context of discovery** (generating theories): Irrational, psychological, creative

- **Context of justification** (testing theories): Rational, logical, rule-governed

Premise 1: TRUE

Popper is absolutely right that there's no algorithm/recursive procedure for:

- Observing planetary orbits $\rightarrow$ inferring $F = Gm_1m_2/r^2$
- Seeing spectral lines $\rightarrow$ inferring atomic structure
- Noting bacterial death near mold $\rightarrow$ inferring penicillin
- Watching apples fall $\rightarrow$ conceptualizing universal gravitation

No recursive rules take you from data to theory.

Premise 2: FALSE

But the implicit premise—that rationality = recursivity—is wrong.

Therefore, the conclusion is wrong: **Scientific discovery is rational, just not recursive.**

Newton's Discovery: Rational But Non-Recursive

Newton observes:

- Planetary orbits (Kepler's laws: ellipses, areas, periods)
- Moon's motion around Earth
- Falling terrestrial objects (apples, stones, etc.)
- Tidal patterns
- Cometary trajectories

Newton infers:

- Universal gravitation—all bodies attract with force proportional to mass and inversely proportional to distance squared: $\mathbf{F = Gm_1m_2/r^2}$

This inference is not recursive because:

- No algorithm goes from "elliptical orbits" to inverse-square law
- No deductive rules connect phenomena to mathematical formula
- No mechanical procedure generates the theory
- No formal inference system captures the leap

But this inference IS rational because:

- It's warranted by the evidence (observations support the theory)
- It unifies diverse phenomena under a single principle
- It's explanatory (shows why planets move as they do)
- It's predictive (makes novel testable predictions)
- It's recognizable as rational by other competent scientists (even if they couldn't have generated it themselves)

This is paradigmatic rationality—just non-recursive rationality.

What Non-Recursive Rationality Looks Like

Non-recursive rational inference involves:

Pattern Recognition:

- Perceptual: Recognizing faces, objects, scenes

- Conceptual: Seeing structural similarity across domains

- Scientific: Recognizing patterns in data (Newton, Darwin, Mendeleev)

Abductive Inference:

- Inference to best explanation

- Not deduction (doesn't guarantee truth)

- Not induction (doesn't just generalize)

- Creative hypothesis generation

Unification:

- Bringing disparate phenomena under single principle

- Recognizing deep structural similarity

- Not mechanical, but insightful

Mathematical Insight:

- Seeing structure, grasping proofs

- Recognizing unprovable truths (Gödel sentence)

- Not derivation, but direct apprehension

Conceptual Innovation:

- Creating new theoretical concepts ("gravitational force," "mass," "field")

- Not combining old concepts recursively

- Genuine conceptual novelty

None of these are recursive. All of them are rational.

Why Popper Was Wrong

Popper concluded that because discovery isn't recursive, it's irrational. But this assumes: **Rational = Recursive**. Once we reject that equation:

- Discovery is non-recursive ✓

- Discovery is rational ✓

- No problem ✓

There IS a "logic of discovery"—just not a recursive logic.

It's the logic of:

- Pattern recognition

- Abductive inference

- Unification

- Mathematical/conceptual insight
- Creative hypothesis generation

All of which are rational but non-recursive.

Other Examples

Darwin: Observing finch beaks, fossils, breeding → inferring natural selection

Mendeleev: Organizing chemical elements → recognizing periodic law

Einstein: Thinking about light, gravity, equivalence principle → developing general relativity **Watson & Crick:** Seeing X-ray crystallography, Chargaff's rules → discovering DNA double helix In every case:

- No recursive algorithm from data to theory
- But paradigmatically rational inference
- Pattern recognition, unification, abductive reasoning

The Upshot

Scientific discovery is:

- Non-recursive ✓
- Rational ✓
- Exemplary of human cognitive capacity ✓

Popper was right that it's non-recursive. Popper was wrong that it's irrational.

His error: Accepting the false equation Rational = Recursive.

Neural Networks Vindicated

The Core Insight

If neural networks were recursive, they couldn't do jackshit.
Because they are non-recursive, there is nothing they cannot do.

Why Recursive Systems Are Limited

Recursive systems (formal proofs, algorithms, rule-based AI) are limited to:

- Following explicit rules
- Deductive inference
- Compositional operations
- What can be specified in advance

They cannot:

- Recognize novel patterns (not covered by rules)
- Make abductive inferences (inference to best explanation)
- Learn from unstructured data
- Adapt to genuinely new situations

This is why Good Old-Fashioned AI (GOFAI) failed:

- Tried to encode knowledge as recursive rules (expert systems)

- Tried to model reasoning as logical inference (theorem provers)

- Assumed intelligence = symbol manipulation via algorithms But human intelligence includes massive non-recursive components:

- Pattern recognition (perceptual, conceptual)

- Concept formation (creating new primitives)

- Semantic grounding (ostension, demonstration)

- Scientific insight (abduction, unification)

- Common sense reasoning (not rule-based)

GOFAI couldn't capture these because it was limited to recursivity.

Why Neural Networks Are Not Limited

Neural networks are not recursive systems. They're based on:

- Matrix operations (not recursive functions)

- Gradient descent (not logical inference)

- Statistical learning (not rule-following)

- Parallel computation (not sequential algorithms)

Therefore, they're not bound by the limitations of recursive systems. They can:

- Learn patterns from data (without explicit rules)

- Approximate both recursive and non-recursive rational capacities

Recursive capacities (syntax):

- Learn grammatical patterns from data

- Generate syntactically well-formed text

- Parse complex sentences

Non-recursive capacities (pattern recognition):

- Recognize faces, objects, scenes

- Identify semantic patterns in text

- Make abductive inferences (suggest plausible explanations)

Semantic approximation:

- Learn word embeddings (semantic similarity)

- Capture conceptual relationships

- Approximate meaning through statistical correlation

Quasi-metalinguistic awareness:

- Generate text about language itself

- Reflect on own outputs (to some degree)

- Simulate understanding of linguistic structure

Key point: They can approximate both recursive and non-recursive rational capacities—unlike purely recursive systems (GOFAI), which can only handle the recursive parts.

The Limitation Question

Do neural networks have limitations? Yes, but not from recursion theory.

Potential limitations include:

- **Embodiment**: Lack of causal connection to world (for language models)

- **Intentionality**: Whether statistical patterns constitute genuine understanding (philosophical question)

- **Grounding**: Whether learned associations capture real reference (semantic question)

But these are different issues from recursion.

And notably:

- Embodied robots with neural networks might overcome embodiment limitations

- Whether these constitute "genuine understanding" is a philosophical question, not a technical one

The point: There's no principled limitation from recursion theory on what neural networks can do.

The Vindication

For decades, critics said:

- "Neural networks are just pattern matchers, not real intelligence"

- "They don't follow logical rules, so they can't really reason"

- "They're not symbolic/recursive, so they can't handle language properly"

Our work shows these criticisms were based on the false equation: Intelligence = Recursivity.

Once we recognize:

- Intelligence includes non-recursive capacities (pattern recognition, concept formation, semantic grounding)

- Much of human rationality is non-recursive (scientific discovery, perception, insight)

Then neural networks being non-recursive is a FEATURE:

- They can approximate the non-recursive parts of intelligence

- They're not limited by the constraints on recursive systems

- They succeeded where rule-based AI failed because they're not recursive

Domains Liberated from Recursive Tyranny

The Pattern

In multiple domains, the false equation **Rational = Recursive** led to:

1. Trying to recursively formalize everything

2. Failing to capture important phenomena

3. Concluding either:

- Those phenomena are irrational/subjective (skeptical response)

- We need more/better recursive formalizations (optimistic response)

But the real problem was the false equation itself.

Once we recognize non-recursive rationality, we can make progress.

Domain 1: Philosophy of Science

The problem:

- GOFAI: Intelligence = recursive symbol manipulation

- Result: Brittle systems, couldn't handle real-world complexity

- Failure by 1990s

The liberation:

- Intelligence includes massive non-recursive components

- Neural networks can approximate both recursive and non-recursive capacities

- Modern AI succeeds by abandoning recursive paradigm

What we can now say:

- AI doesn't need to be recursive to be intelligent

- Statistical/neural approaches are scientifically respectable

- No principled limitation from recursion theory

Domain 2: Epistemology

The problem:

- Foundationalism: Knowledge requires recursive justification chains

- Regress: Justification requires more justification, ad infinitum

- Skepticism: If no foundation, no knowledge

The liberation:

- Perceptual knowledge can be direct (pattern recognition)

- Coherentism viable (mutual support, not linear chains)

- Particularism: Case-by-case judgment, not universal rules

What we can now say:

- Knowledge doesn't require recursive foundations

- Direct perceptual justification is legitimate

- Pattern recognition grounds knowledge

Domain 3: Philosophy of Mind

The problem:

- Computationalism: Mind = recursive computation
- Functionalism: Mental states = computational states
- Struggle: Qualia, intentionality, consciousness resist formalization

The liberation:

- Mental capacities include non-recursive elements
- Pattern recognition is neural but non-algorithmic
- Thought transcends computation

What we can now say:

- Mind not purely computational
- Non-recursive capacities are real and essential
- Neuroscience studies non-recursive processes

Domain 4: Philosophy of Language

The problem:

- Compositional semantics dogma: All meaning is recursive
- Struggle: Ostension, naming, context resist formalization
- Paradoxes: Self-reference, indexicals, demonstratives

The liberation:

- Semantics includes essential non-recursive elements (X)
- Linguistic competence = UG (recursive) + X (non-recursive)
- Full semantics cannot be fully formalized

What we can now say:

- Compositional semantics models fragments, not the whole
- Ostension and naming are essential, non-recursive
- Meta-semantic capacity grounds language

Domain 5: Ethics

The problem:

- Moral reductionism: Ethics needs universal recursive principles
- Utilitarianism, Kantianism: Reduce morality to algorithm
- Struggle: Particular cases resist universal rules

The liberation:

- Moral perception: Seeing salience directly in situations
- Particularism: Case-by-case judgment, not rule-application

- Phronesis (practical wisdom): Trained perception, not algorithm

What we can now say:

- Ethics doesn't need universal recursive principles
- Moral expertise is pattern recognition (trained perception)
- Virtue ethics vindicated

Domain 6: Mathematics

The problem:

- Formalism: Mathematics = symbol manipulation
- Struggle: Mathematical intuition seems irrational
- Wars: Formalists vs. intuitionists

The liberation:

- Mathematical intuition is rational (pattern recognition)
- Understanding proofs requires seeing why steps follow
- Gödel sentence recognizable as true (non-formally)

What we can now say:

- Mathematical knowledge includes non-formal elements
- Intuition is rational pattern recognition
- Understanding transcends formal verification

Domain 7: Artificial Intelligence

The problem:

- GOFAI: Intelligence requires symbolic recursive rules
- Expert systems, logic-based reasoning
- Catastrophic failure by 1990s

The liberation:

- Intelligence includes non-recursive capacities
- Neural networks approximate pattern recognition
- Statistical AI succeeds where symbolic AI failed

What we can now say:

- Intelligence doesn't require recursivity
- Pattern-matching is not inferior—it's essential
- Modern AI vindicated by abandoning recursive paradigm

The Positive Program

Breaking the Stranglehold

Once we reject **Rational = Recursive**, we can make progress in understanding:

- How science actually works (discovery is rational)
- How knowledge is justified (not just recursively)
- How mind works (includes non-computational capacities)
- How language works (semantics includes X)
- How ethics works (not just rule-following)
- How mathematics works (intuition is rational)
- How AI can succeed (doesn't need to be recursive)

What Non-Recursive Rationality Includes

Pattern Recognition:

- Perceptual, conceptual, scientific
- Seeing structures, relationships, similarities
- Foundation of rational thought

Abductive Inference:

- Inference to best explanation
- Creative hypothesis generation
- Scientific discovery mechanism

Semantic Grounding:

- Ostensive reference ("that child there")
- Demonstrative grounding
- Connecting language to world

Concept Formation:

- Creating new semantic primitives
- Not recursive combination
- Genuine innovation

Metalinguistic Awareness:

- Talking about language itself
- $E_0 \rightarrow E_1 \rightarrow E_2$ hierarchy
- Non-recursive semantic activation

Mathematical Insight:

- Seeing why proofs work
- Recognizing unprovable truths

- Direct apprehension of structure

Moral Perception:

- Seeing moral salience directly

- Recognizing particular obligations

- Not applying rules, perceiving features

Aesthetic Judgment:

- Recognizing beauty, harmony, excellence

- Not rule-based (no algorithm for good art)

- But not arbitrary—can be educated, refined

Practical Wisdom (Phronesis):

- Knowing what to do in particular situations

- Not following universal rules

- Case-based, experience-based, judgment

All of these are:

- Non-recursive (not algorithmic, not rule-following)

- Rational (warranted, appropriate, truth-tracking)

- Essential to human cognition

Research Programs

For each domain:

Epistemology:

- Develop non-foundationalist theories

- Understand direct perceptual justification

- Explore particularist, coherentist approaches

Philosophy of Mind:

- Investigate non-computational mental capacities

- Study concept formation, insight, creativity

- Connect to neuroscience of non-recursive processes

Philosophy of Language:

- Recognize meta-semantic capacity (X) as central

- Study ostension, naming, demonstration

- Accept semantic non-compositionality

Ethics:

- Develop particularist moral theories

- Study moral perception and expertise

- Abandon search for recursive moral principles

Mathematics:

- Vindicate mathematical intuition

- Study non-formal aspects of mathematical knowledge

- Connect to cognitive science of mathematical thinking

Artificial Intelligence:

- Embrace neural/statistical approaches

- Stop trying to force everything into recursive molds

- Study how to approximate non-recursive rationality

The Unifying Theme

Rationality is broader than recursivity.

Two modes of rationality:

Recursive Rationality:

- Following explicit rules

- Deductive inference

- Compositional operations

- Algorithmic procedures

- Important but LIMITED

Non-Recursive Rationality:

- Pattern recognition

- Abductive inference

- Semantic grounding

- Concept formation

- Insight, perception, judgment

- Equally important, BROADER domain

Both are rational. Neither reduces to the other.

Summary of Key Claims

Part I: Language Structure

- Syntax is recursive (definitionally—Chomsky was right about this)

- Semantics is not recursive (includes ostension, naming, concept formation, metalinguistic awareness)

- You can use recursive syntax to build expressions with non-recursive semantics (Example: "the totality of expressions of English")

- Linguistic competence = UG (recursive) + X (non-recursive)

- X is essential (naming, ostension, meta-semantic capacity—not mere performance)

Part II: Computational Implications

- Complete computational semantics is impossible (principled, not just practical limitation)

- AI/neural networks are not recursive systems (therefore not constrained by recursion theory)

- Neural networks being non-recursive is favorable (allows approximation of non-recursive rational capacities)

- Modern AI succeeded where GOFAI failed because it's not recursive

Part III: Rationality

- The false equation: **Rational = Recursive** (poisoned multiple domains)

- Scientific discovery is rational but non-recursive (Popper was wrong)

- Non-recursive rationality includes: pattern recognition, abduction, semantic grounding, concept formation, insight, perception, judgment

- Rationality cannot recursively characterize itself (by main theorem)

- Breaking the stranglehold: Recognize non-recursive rationality in science, epistemology, mind, language, ethics, mathematics, AI

Part IV: Liberation

Once we reject **Rational = Recursive**, we can make progress in understanding:

- How science actually works (discovery is rational)

- How knowledge is justified (not just recursively)

- How mind works (includes non-computational capacities)

- How language works (semantics includes X)

- How ethics works (not just rule-following)

- How mathematics works (intuition is rational)

- How AI can succeed (doesn't need to be recursive)

Conclusion

The Arc of the Argument

1. **Main Theorem**: The class of recursive logics is not recursively enumerable

2. **Reframing**: Recursion cannot recursively enumerate itself

3. **Extended to language**: Natural languages are hierarchies ($E_0 \cup E_1 \cup \ldots$), not single recursive classes

4. **Syntax vs. Semantics**: Syntax is recursive, semantics is not (includes non-recursive X)

5. **Linguistic competence**: UG (recursive) + X (non-recursive meta-semantic capacity)

6. **Computational Linguistics**: Can model syntax fully, cannot model full semantics (principled limitation)

7. **AI/Neural Networks**: Not recursive systems, therefore not limited by recursion constraints—favorable!

8. **The false equation: Rational = Recursive** (infected multiple domains)

9. **Scientific discovery**: Rational but non-recursive (Popper was wrong)

10. **Non-recursive rationality**: Pattern recognition, abduction, semantic grounding, concept formation, insight

11. **Breaking the stranglehold**: Recognize non-recursive rationality across all domains

The Upshot

The theorem reveals a deep structural feature of rationality itself: **Rationality cannot recursively characterize its own scope.**

Just as:

- Formalism cannot formalize its own boundaries

- Recursion cannot recursively enumerate itself

- Language cannot linguistically characterize language So

too:

- Rationality transcends recursive specification

This is not a limitation but a liberation: We can stop trying to force everything into recursive molds and recognize the full breadth of rational human capacities—including the non-recursive ones that make science, language, thought, and intelligence actually work.

Transition to Part V

We've broken the stranglehold of recursion by showing that the equation **Rational = Recursive** is false and has caused catastrophic failures across multiple domains.

Now Part V develops the positive account: **Rationality is fundamentally pattern recognition.** This view:

- Dissolves Hume's problem of induction

- Recovers all functions attributed to deduction

- Shows formalization (Euclid, Chomsky) is secondary

- Refutes Wittgenstein's rule-following paradox

- Provides a unified account of rational thought

Rationality is seeing, not calculating.

That's what Part V proves, and that's what completes the revolution.

Part V: Rationality as Pattern Recognition

The Core Reframing

From Calculation to Seeing

Parts I-IV established what rationality is NOT:

- Not purely recursive

- Not exhaustively formalizable

- Not reducible to rule-following

- Not algorithmic computation

Now we establish what rationality IS:

RATIONALITY = PATTERN RECOGNITION

More precisely: **Rationality is the skilled capacity to recognize truth-relevant patterns in situations and respond appropriately to them.**

Key Features

Non-algorithmic: No procedure guarantees it **Trainable:** Improves with practice and feedback **Systematic:** Exhibits recognizable patterns (not random) **Productive:** Enables discovery, not just verification **Holistic:** Considers multiple features together

What It Includes

Pattern recognition (primary):

- Seeing structures, relationships, similarities

- Perceptual, conceptual, scientific

Deductive reasoning (special case):

- Rigid formal patterns

- Maximally explicit, exceptionless

Abductive inference:

- Recognizing explanatory patterns

- Inference to best explanation

Perceptual judgment:

- Seeing what's there

- Direct apprehension

Self-correction:

- Recognizing error patterns

- Adapting when patterns mislead

Why This Matters

The false equation **Rational = Recursive** created pseudo-problems:

- **Hume's problem:** Induction seems irrational (because it's not deductive)

- **Popper's irrationalism:** Scientific discovery can't be rational (no algorithm for it)

- **Wittgenstein's paradox:** Rule-following leads to infinite regress

- **GOFAI's failure:** Intelligence can't be achieved without symbolic rules

- **Foundationalist anxiety**: Knowledge requires impossible recursive justification

All dissolve once we recognize: Rationality is broader than recursivity.

The Definition of Rationality

The Concise Definition

RATIONALITY = The skilled capacity to recognize truth-relevant patterns in situations and respond appropriately to them.

Unpacking

Skilled: Developed through practice, improvable, not innate algorithm

Capacity: An ability, not a set of rules to follow

Recognize: Seeing, perceiving, detecting—not calculating or deriving

Truth-relevant patterns: Structures, similarities, relationships, anomalies that bear on what's true, what explains, what predicts

In situations: Context-dependent, domain-specific, particular cases

Respond appropriately: Act, infer, judge, revise in ways that track truth

What Patterns Are Truth-Relevant?

Truth-relevant patterns include:

Evidential patterns: What supports what
Explanatory patterns: What accounts for what
Coherence patterns: What fits together **Predictive**
patterns: What follows from what

Structural patterns: Underlying similarities across surface differences

Anomaly patterns: What doesn't fit, what's surprising

Causal patterns: What produces what effects

These patterns are objective features of reality and reasoning, not arbitrary constructs.

Rationality as Skill, Not Rule-Following

Compare rationality to:

Carpentry:

- Skilled carpenters don't follow explicit rules

- They SEE what the situation requires

- Respond appropriately to wood grain, load-bearing, aesthetics

- Systematic (not random) but non-algorithmic

Chess mastery:

- Masters don't calculate move-trees algorithmically

- They SEE threats, opportunities, positional patterns
- Respond with moves that fit the pattern
- Trainable through practice, but not rule-based

Medical diagnosis:

- Expert diagnosticians don't apply decision trees mechanically
- They RECOGNIZE disease patterns
- See which symptoms cluster together
- Developed through case experience, not rule memorization

Similarly, rationality:

- Experts don't calculate "apply rule R to situation S"
- They SEE what the situation calls for
- Pattern recognition trained through experience

Why Pattern Recognition Is More Fundamental Than Deduction

Even using deductive rules requires pattern recognition: To
apply modus ponens ("If P then Q; P; therefore Q"):

- You must recognize that your situation matches this pattern
- You must see that you have P and (P→Q)
- You must perceive that Q follows

Each step is pattern recognition.

Therefore: **Deduction presupposes pattern recognition, but not vice versa.**

Pattern recognition is the foundation. Deduction is a derivative tool.

Pattern Recognition and Discovery

The Question

If rationality is fundamentally pattern recognition, what becomes of scientific discovery?

Answer: Pattern recognition doesn't just describe discovery—it enables and explains it.

Newton's Discovery of Universal Gravitation

What happened:

- Observed: Planetary orbits, moon's motion, falling objects, tides, comets
- Recognized: These mechanisms share a structural pattern
- Saw: Inverse-square force law unifies all phenomena

The discovery WAS the pattern recognition:

- Seeing how disparate observations fit together

- Recognizing the mathematical structure underlying diverse phenomena

- Perceiving what this pattern explains and predicts

Not:

- Applying deductive rules

- Following algorithmic procedure

- Deriving from prior theories

Darwin's Discovery of Natural Selection

What happened:

- Observed: Variation in offspring, competition for resources, differential survival, inheritance

- Recognized: These mechanisms, operating over time, generate adaptation without design

- Saw: Selection as a generative process

The discovery WAS the pattern recognition:

- Seeing how simple mechanisms compound

- Recognizing selection as explanatory pattern

- Perceiving what this structure explains

Not:

- Applying deductive rules

- Following algorithmic procedure

- Deriving from prior theories

The General Pattern

Scientific discovery = Recognizing a pattern that unifies disparate phenomena

Other examples:

- **Mendeleev:** Organizing elements → recognizing periodic pattern

- **Kekulé:** Benzene structure → recognizing circular/ring pattern

- **Watson & Crick:** DNA → recognizing double helix pattern

- **Einstein (relativity):** Seeing structural equivalence between acceleration and gravity In
every case:

- Discovery ≠ deriving conclusions from premises

- Discovery = seeing structure that wasn't visible before

- **The seeing IS pattern recognition**

Does Pattern Recognition Mediate Data-Modeling?

YES.

Reality: You must SEE what kind of pattern might be there

- Linear? Exponential? Periodic? Chaotic? Network structure?

- Statistical techniques only work once you've recognized what type of pattern to look for

Example: Planetary motion

Same observational data led to different models:

- **Ptolemy:** Recognized circular motion → epicycles (circles on circles)

- **Kepler:** Recognized elliptical pattern → three laws

- **Newton:** Recognized inverse-square pattern → universal gravitation

Each required seeing a new pattern in the data.

The calculations came AFTER pattern recognition.

Does Pattern Recognition Mediate Theoretical Leaps?

YES.

The theoretical leap is:

- Recognizing that hypothesized structure would generate observed patterns

- Then recognizing additional patterns this predicts

This is pattern recognition at multiple levels:

- Patterns in data

- Patterns explained by mechanism

- Patterns predicted by mechanism

The Mechanisms of Pattern Recognition

How does pattern recognition actually work?

Similarity detection:

- Seeing what's relevantly alike despite surface differences

- Newton: planetary orbits ~ falling apples (both attracted to mass)

- Darwin: artificial selection ~ natural processes (both shape populations)

Structural mapping:

- Seeing how structure in one domain mirrors another

- Kekulé: snake circle → molecular ring

- Rutherford: solar system → atomic structure

Anomaly detection:

- Seeing what doesn't fit current pattern

- Mercury's orbit didn't fit Newton → Einstein

- Photoelectric effect didn't fit Maxwell → quantum mechanics

Chunking/abstraction:

- Seeing complex phenomena as instances of simpler patterns

- Chess masters see "king-side attack" not just positions

- Physicists see "conservation law" not just equations

Unification:

- Seeing multiple patterns as instances of deeper pattern

- Maxwell: electricity + magnetism = electromagnetism

- Weinberg-Salam: electromagnetic + weak = electroweak

Is This Trainable?

YES. Pattern recognition is:

Neural: Implemented in brain circuitry **Plastic:**
Modifiable through experience **Reinforceable:**
Strengthened through feedback

Transferable: Patterns learned in one context help in others

Training methods:

- Immersion: Expose to many examples in domain

- Variation: Show same pattern in different contexts

- Contrast: Show similar-but-different patterns

- Guided attention: Point out relevant features

- Practice: Give opportunities to recognize patterns

- Feedback: Correct misrecognition, reinforce correct recognition

This is how we ACTUALLY teach:

- Medical diagnosis (cases, not rules)

- Scientific reasoning (exemplars, not algorithms)

- Mathematical thinking (worked examples, not just proofs)

- Expert judgment (apprenticeship, not rulebooks)

The Upshot

Pattern recognition doesn't just describe discovery—it enables and explains it.

Discoveries happen when someone recognizes a pattern others missed. This
can be:

- Trained (through appropriate practice)

- Improved (through feedback and reflection)

- Explained post-hoc (by identifying what pattern was recognized)

But not:

- Algorithmized (reduced to explicit rules)

- Guaranteed (applied mechanically)

- Replaced by deduction (which comes after, for verification)

Dissolution of Hume's Problem of Induction

The Problem

We observe: The sun has risen every day in the past. We
infer: The sun will rise tomorrow.

Hume asks: What justifies this inference?

Attempted justification: "The future will resemble the past" (Uniformity of Nature)

Hume's challenge: How do you justify THAT principle?

- Not by logic (future being different isn't contradictory)

- Not by experience (that's circular—using induction to justify induction)

Hume's conclusion: Induction has no rational justification. It's mere habit, custom, psychological expectation

—not reason.

Why It Seemed Intractable

250 years of failed attempts to "solve" Hume's problem:

- Probabilistic justification (Bayes)

- Pragmatic justification (Reichenbach)

- Transcendental justification (Kant)

- Natural kind justification (Mill)

All failed because the problem was misconceived.

Because it presupposed: Rationality = Deduction

If rationality = following rules that guarantee truth-preservation:

- Induction isn't rational (doesn't guarantee truth)

- Need some RULE that justifies it

- No such rule exists without circularity

The Solution: Pattern Recognition View

The problem dissolves. Hume's "riddle" was a pseudo-problem.

Why:

Induction IS pattern recognition.

When you observe:

- Sun rose yesterday, day before, day before that...

And infer:

- Sun will rise tomorrow

You're:

- Recognizing a pattern (regular recurrence)

- Seeing structural stability (astronomical mechanism)

- Responding appropriately (expecting continuation)

This IS rationality. It doesn't need further justification.

KEY MOVE: Pattern Recognition Is Basic

Pattern recognition is basic—it's not justified by something else, it's the foundation.

OLD VIEW (Rationality = Deduction):

- Deduction is basic (self-justifying)

- Induction needs justification

- Since no justification exists → induction is irrational

NEW VIEW (Rationality = Pattern Recognition):

- Pattern recognition is basic (self-justifying)

- Induction IS pattern recognition

- Therefore induction IS rational (no further justification needed)

Why Pattern Recognition Doesn't Need External Justification

It's constitutive of rationality:

- Asking "Why is pattern recognition rational?" is like asking "Why is seeing visual?"

- Pattern recognition DEFINES what rationality is

It's self-correcting:

- When patterns mislead, we recognize that too (anomaly detection)

- The system corrects itself through further pattern recognition

It's truth-tracking:

- Pattern recognition evolved/developed because it tracks real patterns

- It works because the world has structure

It's more fundamental than deduction:

- Even applying deductive rules requires recognizing patterns

- Recognizing that THIS is an instance of modus ponens = pattern recognition

The Circularity Objection Dissolves

Hume said: "You can't use induction to justify induction—that's circular!"

Response: Right! Because induction doesn't need justification by something else. It's basic.

Analogy:

Wittgenstein's worry (applied to vision):

- To see a circle, you must interpret visual input

- Interpreting requires following interpretation-rules

- Following those requires interpreting them

- Regress...

Obviously wrong: Seeing is DIRECT. No interpretation needed.

Similarly: Pattern recognition is DIRECT. No recursive justification needed.

What About "The Future Will Resemble the Past"?

This isn't a PRINCIPLE we need to justify. It's a PATTERN we recognize directly.

You SEE stability, regularity, persistence in phenomena.

That's not an inference from a principle—it's direct pattern recognition.

Sometimes the pattern breaks (sun fails to rise = Earth stops rotating). Then you recognize THAT pattern (anomaly) and adjust.

Goodman's "New Riddle": Grue

Goodman's challenge:

- Why project "green" rather than "grue" (green until time t, then blue)?

Pattern recognition answer:

- We recognize "green" as a natural pattern (stable property)

- "Grue" is gerrymandered (defined by arbitrary time t)

- Pattern recognition tracks natural joints, not arbitrary constructions

- This is objective: "green" corresponds to wavelength; "grue" doesn't

Why Hume Got It Wrong

Hume assumed: Rationality = deduction/logic

So when he saw that induction isn't deductively valid, he concluded it's not rational.

But he was using the wrong standard.

Induction is rational by the correct standard (pattern recognition), just not by the incorrect standard (deduction).

Hume himself almost saw this!

He said induction is "custom" or "habit"—i.e., our cognitive nature.

He was RIGHT that it's basic to cognition.

He was WRONG to think that makes it non-rational.

It's rational BECAUSE it's basic to cognition. Pattern recognition is what rational agents DO.

Summary: Hume's Riddle Dissolved

OLD VIEW:

- Rationality = deduction
- Induction ≠ deduction
- Therefore: induction needs justification
- No justification exists
- Therefore: induction is irrational (mere habit)

NEW VIEW:

- Rationality = pattern recognition
- Induction = pattern recognition
- Therefore: induction IS rational (no further justification needed)
- Pattern recognition is basic, self-correcting, truth-tracking
- Hume's "riddle" was a pseudo-problem from wrong conception of rationality

One-line answer: Hume's riddle dissolves once we see that induction doesn't need justification by deduction, because rationality is pattern recognition, and induction IS pattern recognition—which is basic, self-correcting, and constitutive of rational thought.

Pattern Recognition Recovers Deduction's Functions

The Question

If rationality is fundamentally pattern recognition, what becomes of deduction?

Does pattern recognition merely add to deduction, or does it subsume what deduction was supposed to do?

Answer: Pattern recognition recovers ALL functions attributed to deduction. Deduction is a special case.

The Seven Functions of Deduction

Traditional view held that deduction provides:

1. Truth-preservation (validity)

2. Consistency-checking

3. Drawing consequences

4. Providing justification

5. Ensuring rigor

6. Transmitting knowledge/certainty

7. Clarifying concepts

We'll show pattern recognition does ALL of these.

Function 1: Ensuring Validity / Truth-Preservation

What deduction was supposed to do:

- Guarantee truth of conclusion from premises
- Valid inference preserves truth mechanically

What actually happens:

- Recognizing valid inference = seeing that conclusion follows from premises
- This is pattern recognition of inferential structure

Evidence:

- People make mistakes in deductive logic (fail to recognize valid patterns)
- Skill improves with practice (pattern recognition is trainable)
- Context matters (same logical form harder to recognize in some contents)

This is pattern recognition, not mechanical rule-following.

Deeper point:

- Checking whether inference is valid requires seeing the pattern
- Computer verification only works because programmers recognized what to check
- At base: pattern recognition of truth-preserving structure

Mathematical proof:

- Formalists say: proof = mechanical symbol manipulation
- Reality: understanding proof = recognizing WHY each step follows
- This is seeing patterns (structural relationships, dependencies, necessities)

Upshot: Truth-preservation is recognized, not mechanically executed.

Function 2: Checking Consistency / Detecting Contradictions

What deduction was supposed to do:

- Detect when statements contradict
- Ensure coherence via formal rules

What actually happens:

- Recognizing contradiction = seeing that statements can't both be true
- Pattern recognition of incompatibility

Evidence:

- Contradictions can be hidden (people miss them)
- Expert logicians better at spotting (trained pattern recognition)
- Sometimes requires insight to see inconsistency

Mathematical case:

- Proving system consistency: requires seeing that no derivation could yield P and $\neg P$
- Often requires insight (recognizing structural properties)
- Example: Euclidean geometry consistent (give model—you recognize axioms all satisfied)

Upshot: Consistency-checking is pattern recognition of compatibility.

Function 3: Making Predictions / Drawing Consequences

What deduction was supposed to do:

- Derive conclusions from premises
- Follow logical rules mechanically

What actually happens:

- Drawing consequences = recognizing what follows from what

Scientific case:

- Newton's laws + initial conditions $\rightarrow$ predict positions
- Looks deductive. But actually:
    - Must recognize which law applies
    - Must see how to set up equations
    - Must perceive what solution means physically
- Each step requires pattern recognition

Mathematical case:

- Theorem: "If function is differentiable, it's continuous"
- Applying: Must recognize THIS function is differentiable, must see theorem applies HERE
- All pattern recognition

Upshot: Drawing consequences is recognizing what-follows-from-what patterns.

Function 4: Providing Justification / Giving Reasons

What deduction was supposed to do:

- Justify beliefs by deriving from other beliefs
- Build chains of justification

What actually happens:

- Justification = showing why conclusion is warranted
- This is showing the evidential/explanatory pattern

Evidence:

- Good justifications make you SEE why conclusion follows
- Bad justifications feel mechanical, don't provide insight
- Understanding justification $\neq$ checking steps mechanically

Upshot: Justification is making patterns visible, not mechanical derivation.

Function 5: Ensuring Rigor / Avoiding Error

What deduction was supposed to do:

- Follow strict rules to avoid mistakes

* Mechanical checking ensures correctness

What actually happens:

* Rigor = careful attention to what follows

* Pattern recognition of inferential structure, gaps, assumptions

Proof-checking:

* Automated: computers verify mechanically

* Human: we see whether proof correct (pattern recognition of validity)

* But even automated checking was set up by humans who recognized what patterns to check

Deeper point:

* "Rigor" isn't just mechanical rule-following

* It's: Recognizing what needs showing, seeing gaps in reasoning, detecting hidden assumptions, perceiving where argument might fail

* All pattern recognition (of inferential structure, weaknesses, dependencies)

Upshot: Rigor is disciplined pattern recognition, not mere rule-following.

Function 6: Transmitting Certainty / Preserving Knowledge

What deduction was supposed to do:

* Transfer certainty from premises to conclusion

* Knowledge transmitted via valid inference

What actually happens:

* Knowledge transmission = recognizing that evidence supports conclusion

* Seeing why conclusion is warranted given what you know

Evidence:

* Can follow valid argument without understanding (just checking steps)

* Real understanding = seeing WHY conclusion follows

* This is pattern recognition

Upshot: Knowledge transmission requires recognizing evidential patterns.

Function 7: Clarifying Concepts / Making Definitions Explicit

What deduction was supposed to do:

* Analyze: Break complex concepts into simpler components

* Define: Give necessary and sufficient conditions

* Clarify: Remove ambiguity through formal precision

What actually happens:

* Conceptual clarification is pattern recognition of meaning-relationships

- Defining "bachelor" as "unmarried adult male": Recognize semantic equivalence, see these capture same pattern, perceive no counterexamples

This is pattern recognition, not deduction:

- Definitions aren't derived; they're recognized as apt

- Good definitions capture patterns of use

- Testing requires seeing whether they fit cases

Deeper point:

- Concept formation (creating NEW concepts) clearly non-deductive:

    - "Natural selection" (Darwin)

    - "Unconscious" (Freud)

    - "Information" (Shannon)

- Not defined from prior concepts—created by recognizing new patterns

Upshot: Pattern recognition recovers conceptual work: recognizing semantic relationships, recognizing meaning-patterns, creating new concepts.

What Deduction Actually Is

Given all the above:

DEDUCTION IS NOT A SEPARATE FACULTY OR METHOD.

Deduction is a SPECIAL CASE of pattern recognition:

- Patterns that are maximally rigid

- Context-independent

- Exceptionless

- Formally codifiable

Characteristics of "deductive" patterns:

- **Necessity:** Pattern admits no exceptions (modus ponens always valid)

- **Formality:** Pattern depends only on logical form, not content

- **Explicitness:** Pattern can be stated as explicit rule

- **Mechanizability:** Pattern can be checked algorithmically (in principle) But
even recognizing and applying these patterns requires pattern recognition.

The Recovery Thesis

Pattern recognition recovers ALL functions attributed to deduction:

1. Truth-preservation (recognizing valid inference)

2. Consistency-checking (recognizing compatibility)

3. Consequence-drawing (recognizing what follows)

4. Justification (showing evidential patterns)

5. Rigor (disciplined pattern recognition)

6. Knowledge-transmission (recognizing support)

7. Conceptual clarity (recognizing meaning-patterns)

In every case:

- Pattern recognition does the actual cognitive work

- Deduction describes a special class of patterns (necessary, formal, explicit)

- But recognizing, applying, and verifying deduction requires pattern recognition

The Radical Claim

Deduction doesn't do ANY work that pattern recognition doesn't do more fundamentally.

Even in paradigm deductive domains:

Mathematics:

- Mathematical intuition is rational (not mystical)

- Recognizing unprovable truths (Gödel sentence)

- Understanding proofs requires seeing why steps follow

- Pattern recognition, not just formal manipulation

Logic:

- Valid inference forms = codified patterns

- But recognizing instances requires pattern recognition

- Creating new logics requires seeing new patterns

Computer science:

- Algorithms = explicit recursive patterns

- But designing algorithms requires recognizing problem-patterns

- Debugging requires anomaly-detection

Why This Matters

Deduction gets demoted:

- Not essence of rationality

- Not foundation of knowledge

- Just one tool among many (useful in formal contexts)

Pattern recognition gets promoted:

- Actual mechanism of rational thought

- Operates in deduction, induction, abduction, perception

- Subsumes what deduction was supposed to do

The theoretical landscape shifts:

OLD:

- Deduction (primary, foundational)

- Induction (secondary, problematic)

- Everything else (non-rational)

NEW:

- Pattern recognition (primary, foundational, rational)

- Deduction (special case: rigid formal patterns)

- Induction (special case: regularity patterns)

- Abduction (special case: explanatory patterns)

- All are modes of pattern recognition

Part V: Rationality as Pattern Recognition (Continued)

Do You Need Euclid? Do You Need Chomsky?

The Question More Succinctly

If you have pattern recognition, do you need Euclid?

If you have pattern recognition, do you need Chomsky?

Do You Need Euclid?

NO.

What Euclid did:

- Made geometric reasoning explicit (axioms, definitions, proofs)

- Made it teachable (via formal steps)

- Made it verifiable (check each step)

- Made it transmissible (write down exactly)

But the actual geometric insight:

- Seeing that triangles with equal sides have equal angles

- Recognizing that parallel lines never meet

- Perceiving that angles in triangle sum to 180°

This is pattern recognition.

Ancient builders had this capacity before Euclid:

- Could see geometric relationships

- Recognize spatial patterns

- Apply geometric principles

- Solve practical problems

Euclid didn't give them rationality. He formalized rationality they already had.

The value of Euclid:

- **Systematization:** Organized scattered insights

- **Rigor:** Made implicit reasoning explicit and checkable

- **Pedagogy:** Created teachable method

- **Extension:** Formal system enables deriving new results

But:

- Ancient architects built before Euclid

- Egyptian pyramids show geometric mastery

- Geometric intuition predates formalization

- Pattern recognition suffices for geometric thinking

So: Do you NEED Euclid?

For what?

- For geometric insight? No. Pattern recognition suffices.

- For teaching efficiently? Yes. Formalization helps.

- For rigorous proof? Yes. Explicit axiomatization enables verification.

- For doing geometry at all? No. People did geometry before and without Euclid.

Analogy:

- Musical notation formalizes music

- Makes it transmissible, analyzable, teachable

- But musicians had music before notation

- Pattern recognition (hearing, creating melodies) is primary

- Notation is secondary (useful, but not necessary for musicality)

Do You Need Chomsky?

This is more complicated because Chomsky claims to describe how people actually think (regarding language). Three possible positions:

1. UG is real but implemented via pattern recognition

2. UG is wrong (pattern recognition all the way down)

3. Hybrid: UG exists for syntax, but X (pattern recognition) does semantic work

We adopt positions 1 AND 3 together.

Position 1: UG Is Real But Pattern Recognition Is the Mechanism

The view:

- Chomsky RIGHT that there's innate linguistic structure

- Chomsky RIGHT that syntax is recursive/generative

- But mechanism isn't "applying rules"—it's recognizing and producing syntactic patterns

- UG = innate pattern-recognition capacity specifically for syntactic structures

Not: "Mind contains rules like 'S → NP + VP'"

But: "Mind structured to recognize NP-VP patterns, embedding patterns, agreement patterns"

How this works:

Children acquire language by:

- Recognizing syntactic patterns in input (not explicitly taught)

- Generalizing those patterns (productive use)

- Patterns they can recognize are constrained by UG (innate structure)

The mechanism:

- UG isn't rule-system that gets applied

- UG is structured pattern-recognition capacity

- Like: Visual system structured to recognize edges, shapes, faces

- Not "rules for edge-detection"

- But neural architecture that naturally responds to edge-patterns

Similarly:

- "S → NP + VP" (real rule)

- Speakers don't calculate "combine noun phrase with verb phrase"

- They SEE/PRODUCE sentence patterns

- The rule describes the pattern they instantiate

Key distinction:

- **Chomsky claims:** "Language is computation—mind applies recursive rules"

- **Pattern recognition view:** "Language uses recursive PATTERNS, recognized by innately structured pattern-recognition capacity (UG), plus non-recursive semantic capacity (X)"

Difference:

- **Chomsky:** Mind has RULES it APPLIES

- **Pattern view:** Mind has SENSITIVITY TO PATTERNS it RECOGNIZES

Why this matters:

If Chomsky right:

- Language is algorithmic

- Should be fully formalizable

- AI should succeed via symbolic rule-systems (GOFAI) If

pattern view right:

- Language is recognitional

- Only partially formalizable (syntax yes, full semantics no)

- AI succeeds via pattern recognition (neural networks)

Empirical evidence favors pattern view:

- Neural networks succeed at language

- GOFAI failed

- Native speakers can't articulate rules they supposedly follow

- Linguistic theories constantly change

Position 3: UG + X (The Hybrid)

This is the synthesis from our earlier analysis:

LINGUISTIC COMPETENCE = UG (recursive syntax) + X (non-recursive semantics)
UG (Universal Grammar):

- Innate capacity for recursive syntactic patterns

- Phrase structure, movement, binding, c-command

- Chomsky was RIGHT about this part

X (the non-recursive supplement):

- Ostensive reference ("that child there")

- Naming capacity ("Let's call him 'Smith'")

- Concept formation (creating new semantic primitives)

- Metalinguistic awareness ($E_0 \rightarrow E_1 \rightarrow E_2$ hierarchy)

- Semantic grounding (connecting language to world)

Crucially:

- UG is recursive/computational

- X is non-recursive/pattern-recognition

- Both are necessary. Neither alone suffices.

The Architecture:

UG (Universal Grammar):

- Innate syntactic structure

- Recursive/compositional capacity

- Universal constraints across languages

- Specialized for syntax

- Enables infinite generativity from finite means

Operates via:

- Innately structured pattern recognition for syntactic patterns
- Recognizing phrase structure, embedding, dependencies
- Producing grammatical structures

X (Meta-semantic capacity):

- Ostensive definition: "That child there" $\rightarrow$ grounding reference perceptually
- Naming capacity: "Let's call him 'Smith'" $\rightarrow$ creating new primitive referring terms
- Concept formation: Creating new semantic atoms
- Metalinguistic awareness: Talking about language itself ($E_0 \rightarrow E_1$)
- Semantic grounding: Connecting language to world via perception, social practice

All of these are:

- Non-recursive (no algorithm generates them)
- Essential to full linguistic competence
- Irreducible to UG
- Pattern recognition (in broad sense—recognizing/creating semantic relationships)

Why Both Are Necessary

UG alone gives you:

- Syntactic structure
- Infinite generativity
- Well-formed expressions

But:

- No way to connect to world
- No new concepts
- No adaptation to novel situations

X alone gives you:

- Semantic grounding
- Reference to world
- Concept formation

But:

- No systematic structure
- No infinite expressiveness
- Limited to finite learned associations

Together:

- UG provides syntactic scaffold
- X fills it with semantic content
- Result: meaningful, well-formed, infinitely expressive language

So: Do You Need Chomsky?

Depends on what "Chomsky" means:

If "Chomsky" = there exists innate linguistic structure (UG):

- YES, you need this
- Pattern recognition alone (blank slate) can't explain:
    - Poverty of stimulus
    - Universal constraints
    - Critical period
    - Systematic acquisition patterns

If "Chomsky" = language is recursive rule-application:

- NO, you don't need this
- The mechanism is pattern recognition (structured by UG), not rule-following

The synthesis:

- UG (Chomsky's insight) + X (critics' insight)
- Recursive (formal) + Non-recursive (grounded)
- Innate structure (UG) + Developed capacity (X through practice)
- Pattern recognition mediates BOTH

The General Point

Pattern recognition is PRIMARY. Formalization (Euclid, Chomsky, Aristotle, Frege) is SECONDARY.

The relationship:

- Formalization doesn't replace pattern recognition—it codifies some patterns that pattern recognition already handles
- Useful for: teaching, verification, systematization, extension
- Not necessary for: having the capacity, exercising it, being rational

Rules as Patterns: Refuting Wittgenstein

The Reconciliation: Rule-Application IS Pattern Instantiation

KEY INSIGHT:

When the mind applies "S $\to$ NP + VP", it's not:

- Mechanically executing an algorithm

- Consulting an internal rulebook

- Following a procedure blindly

It's:

- Recognizing the S-pattern (this is where a sentence goes)

- Instantiating NP-VP structure (producing/recognizing this pattern)

- Seeing that this matches the rule (pattern recognition that this satisfies the constraint)

Not contradictory:

- Rules DESCRIBE patterns

- Applying rules = instantiating patterns

- Following rules = producing instances that match the pattern

Analogy: Musical Composition

Musical composition in sonata form:

- There are RULES (exposition, development, recapitulation)

- Composers FOLLOW them (produce sonatas)

- But mechanism is: recognizing and instantiating the sonata-pattern

- Not: mechanically checking "Did I do exposition? Yes. Now development..."

- But: seeing/creating structure that fits the pattern

RULES and PATTERNS are two ways of describing the same structure:

- Rules: explicit, formal, prescriptive ("Do this")

- Patterns: implicit, recognizable, descriptive ("Like this")

The mind works with PATTERNS. Linguists describe those patterns as RULES. But the patterns are real

—they're what the mind recognizes and instantiates.

Chess example:

- Knight moves in L-shape (real rule)

- Chess masters don't calculate "two squares, then one perpendicular"

- They SEE knight-move patterns

- The rule describes the pattern they recognize

Wittgenstein's Rule-Following Paradox

THE REGRESS ARGUMENT:

To follow rule R, you need to interpret R

Interpreting R requires following interpretation-rule R'
Following R' requires interpreting R'

This requires rule R''

Infinite regress...

Wittgenstein's conclusion:

Rule-following can't be intellectualist (based on interpretation/understanding).

It must be: "Just doing it"—blind conformity to training, no principled basis except agreement with community practice.

As stated: This is wrong.

The Refutation via Pattern Recognition

THE KEY MOVE:

Rule-following doesn't require interpretation when the rule describes a pattern you can directly recognize. No regress because:

- You follow rule R by recognizing/instantiating the R-pattern
- Recognizing patterns doesn't require following an interpretation-rule
- Pattern recognition is BASIC—not mediated by further rules
- Therefore: no regress

Detailed:

Wittgenstein assumes:

- Following rule R = interpreting R, then acting on interpretation
- Interpretation = applying another rule (R')
- Hence regress

But actually:

- Following rule R = instantiating the R-pattern
- Pattern instantiation = direct recognition + production
- No interpretation needed (no R' required)
- Hence no regress

Example: Adding 2

Wittgenstein's puzzle:

- Rule: "Add 2"
- How do you know to continue: 1000, 1002, 1004... (not 1000, 1004, 1008...)?
- Any finite sequence compatible with infinitely many rules
- So how do you know which rule you're following?

Wittgenstein's answer:

- You DON'T know (no fact determines which rule)
- You just DO what training inclined you to do

- Agreement with community makes it "correct"

Pattern recognition answer:

- You directly recognize the add-2 pattern

- The pattern is: each number is 2 more than previous

- This is directly perceived, not inferred from interpretation-rule

- 1000, 1004, 1008 doesn't instantiate add-2 pattern (violates at step 2: should be 1002, not 1004)

- The pattern itself determines correctness (objective, not social)

Why Pattern Recognition Blocks the Regress

The regress requires:

- Each step mediated by interpretation

- Interpretation requires another rule

- That rule requires interpretation

- Etc.

Pattern recognition breaks this:

- Recognition is DIRECT (not mediated by interpretation)

- No "rule for recognizing patterns" needed

- Pattern recognition is GROUND FLOOR (can't regress further)

Analogy:

Obviously wrong: "To see a circle, you must follow circle-recognition rules"

- Seeing a circle is DIRECT

- Not mediated by interpretation

- Pattern recognition in visual system

- No regress

Similarly (for rule-following):

- Following "S → NP + VP" is recognizing/instantiating syntactic pattern

- Direct (not mediated by interpretation)

- Pattern recognition in linguistic system

- No regress

What Wittgenstein Got Wrong

He assumed:

- "Principled rule-following" requires:

    - Explicit interpretation

    - Intellectual grasp of rule's meaning

- Mediation by conscious understanding

Since that leads to regress, he concluded:

- No principled rule-following

- Just: blind conformity to training

- Community agreement is only "correctness"

But he missed:

- Principled rule-following can be:

    - Direct pattern recognition/instantiation

    - Skilled, not intellectual

    - Non-interpretive, but still principled

    - Normative (can be done correctly/incorrectly), not just communal

The Normativity Problem

Wittgenstein's worry:

- If rule-following is "doing what comes naturally after training," where does normativity come from?

- How can there be RIGHT/WRONG ways to follow a rule?

His answer:

- Normativity = community agreement

- "Correct" = conforms to what community does

- No objective standard beyond social practice

Pattern recognition answer:

- Normativity comes from the pattern itself

- Correct = actually instantiates the pattern

- Incorrect = produces different pattern

- Can be determined by checking instance against pattern

- Not: determined by polling community

Example:

- Rule: "Add 2"

- Correct: 2, 4, 6, 8, 10...

- Incorrect: 2, 4, 6, 9, 11...

Why is second incorrect?

- Wittgenstein: Because community wouldn't agree

- Pattern recognition: Because it doesn't instantiate add-2 pattern (violates at step 4: should be 8, not 9)

The pattern itself provides the normative standard.

Not social agreement. Not interpretation. The pattern is REAL—recognizing/instantiating it correctly matters objectively.

The Positive Account of Rule-Following

How rule-following works (via pattern recognition):

1. Rules describe patterns

- "Add 2" describes: each number is 2 more than previous
- "S $\rightarrow$ NP + VP" describes: sentence structure = noun-phrase + verb-phrase

2. Following rule = instantiating pattern

- When you follow "add 2," you produce sequence matching add-2 pattern
- When you follow "S $\rightarrow$ NP + VP," you produce structure matching NP+VP pattern

3. Recognition is direct (no interpretation needed)

- You SEE add-2 pattern (not by applying interpretation-rule)
- You SEE NP+VP pattern (not by calculating)
- Pattern recognition is BASIC cognitive capacity

4. Correctness is objective (not just social)

- Correct = actually instantiates the pattern
- Incorrect = produces different pattern
- Can be determined by checking instance against pattern
- Not: determined by polling community

5. Learning is training pattern recognition

- Through examples, feedback, practice
- Develops capacity to recognize/produce pattern
- Not: memorizing explicit interpretation-rules

6. No regress

- Because recognition is direct (not rule-mediated)
- Pattern recognition is ground floor
- Can't ask "What rule tells you how to recognize patterns?" (category mistake)

Summary: Wittgenstein Refuted

YES:

- The mind DOES apply rules like "S $\rightarrow$ NP + VP"
- Real cognitive structures
- Not just linguist's descriptions
- Actual patterns the mind instantiates

Rule-application = pattern instantiation

- Not contradiction

- Rules describe patterns

- Applying rules = recognizing/producing instances of those patterns

This REFUTES Wittgenstein:

- Rule-following doesn't require interpretation

- Pattern recognition is DIRECT (no regress)

- Normativity from patterns themselves (not social agreement)

- Principled, not "wiggling and jiggling"

Wittgenstein was wrong:

- Showed interpretive rule-following fails (correct)

- Concluded no principled rule-following exists (wrong)

- Missed: recognitional rule-following via pattern recognition

The truth:

- Rules are real (Chomsky right)

- Mechanism is pattern recognition (our view right)

- Both together: rules = patterns; following = instantiating; recognition = basic

Synthesis and Implications

The Core Position

RATIONALITY = PATTERN RECOGNITION

Concise definition: Rationality is the skilled capacity to recognize truth-relevant patterns in situations and respond appropriately.

Key features:

- Non-algorithmic (no procedure guarantees it)

- Trainable (improves with practice)

- Systematic (not random)

- Productive (enables discovery)

What This Replaces

OLD VIEW: Rationality = Deduction

- Following explicit logical rules

- Step-by-step derivation

- Algorithmic, recursive

- Paradigm: Mathematical proof, formal logic

NEW VIEW: Rationality = Pattern Recognition

- Seeing structures, similarities, relationships

- Holistic judgment

- Non-algorithmic but systematic

- Paradigm: Scientific discovery, expert diagnosis, skilled perception

The Major Results

Hume's Problem Dissolves:

- Induction doesn't need justification by deduction

- Induction IS pattern recognition (basic, self-justifying)

- No regress, no circularity problem

Pattern Recognition Recovers All Deductive Functions:

- Truth-preservation, consistency-checking, consequence-drawing

- Justification, rigor, knowledge-transmission, conceptual clarity

- Deduction is special case (rigid formal patterns)

- But recognizing/applying deduction requires pattern recognition

Formalization Is Secondary:

- Euclid codified geometric patterns humans already recognized

- Chomsky identified linguistic patterns (UG) but misidentified mechanism

- Formalization useful but not necessary for rational capacity

Linguistic Competence = UG + X:

- UG (syntax): Recursive, innate, operates via structured pattern recognition

- X (semantics): Non-recursive, includes ostension, naming, concept formation, metalinguistic awareness

- Both necessary; neither alone suffices

Wittgenstein's Paradox Refuted:

- Rule-following = pattern instantiation (not interpretation)

- Pattern recognition is direct (no regress)

- Normativity from patterns themselves (objective, not social agreement)

- Principled rule-following via recognition, not "just doing it"

Implications Across Domains

Philosophy of Science:

- Discovery is rational (not mysterious/irrational)

- Scientific reasoning is pattern recognition

- No algorithm for discovery, but not arbitrary

- Trainable, systematic, objective

Epistemology:

- Knowledge doesn't require recursive foundations

- Perceptual knowledge is direct (pattern recognition)

- Justification can be non-recursive

- Coherence is rational without being algorithmic

Philosophy of Mind:

- Thinking includes non-computational capacities

- Pattern recognition is physical but non-algorithmic

- Neural networks vindicated (they do pattern recognition)

- Mind not purely computational

Philosophy of Language:

- Semantics not fully compositional

- Includes essential non-recursive elements (X)

- Formal semantics limited to fragments

- Full competence requires UG + X

Ethics:

- Moral reasoning can be rational without universal rules

- Moral perception (recognizing moral salience)

- Particularism viable (case-by-case judgment)

- Virtue ethics vindicated (character-based, not rule-based)

Mathematics:

- Mathematical intuition is rational

- Understanding proofs requires seeing why steps follow

- Gödel sentence recognizable as true (non-formally)

- Mathematical knowledge includes non-formal elements

Artificial Intelligence:

- Neural networks succeed because non-recursive

- Can approximate both recursive and non-recursive rationality

- GOFAI failed because limited to recursion

- No principled limitation from recursion theory

Theoretical Landscape Shift

OLD:

- Deduction (primary, foundational, rational)
- Induction (secondary, problematic, questionable)
- Everything else (non-rational, mere psychology)

NEW:

- Pattern recognition (primary, foundational, rational)
- Deduction (special case: rigid formal patterns)
- Induction (special case: regularity patterns)
- Abduction (special case: explanatory patterns)
- All are rational modes of pattern recognition

Pedagogical Implications

Teaching rationality:

OLD METHOD:

- Teach logic rules
- Drill on valid/invalid forms
- Test whether can apply modus ponens

NEW METHOD:

- Expose to paradigm cases
- Train pattern recognition through examples
- Develop judgment through practice
- Logic rules are one tool, not foundation

Assessing rationality:

OLD CRITERION:

- Did they follow the rules?
- Are they logically consistent?
- Can they construct valid arguments?

NEW CRITERION:

- Do they see relevant patterns?
- Do they track truth reliably?
- Do they respond appropriately to evidence?
- Can they recognize when they're wrong?

The Positive Program

Research directions:

For each domain:

- Identify truth-relevant patterns specific to that domain
- Study how experts recognize those patterns
- Develop training methods to improve pattern recognition
- Understand why certain patterns are truth-relevant
- Recognize limits of recursive formalization

Pattern recognition is:

- Studyable (via cognitive science, neuroscience, philosophy)
- Trainable (via appropriate practice)
- Explainable (via identifying what patterns were recognized)
- But not algorithmizable (no complete recursive specification)

Final Formulation

The false equation that infected multiple domains: **RATIONAL = RECURSIVE Led to:**

- Hume's problem (induction seems irrational)
- Popper's irrationalism (discovery can't be rational)
- Wittgenstein's paradox (rule-following impossible)
- GOFAI's failure (intelligence needs symbolic rules)
- Foundationalist anxiety (knowledge needs recursive justification)

The correction: RATIONALITY IS BROADER THAN RECURSIVITY

Two modes of rationality:

- **Recursive:** Following explicit rules, deductive, algorithmic (special case)
- **Non-Recursive:** Pattern recognition, abductive, insight-based (general case)

"Rationality is seeing, not calculating."

Or more precisely: "Rationality is primarily pattern recognition; calculation is a special case useful in certain domains."

Conclusion

We Have Established

1. **Rationality = pattern recognition** (skilled capacity to recognize truth-relevant patterns and respond appropriately)

2. **Pattern recognition enables discovery** (mediates data-modeling, theoretical leaps, insights—it's the mechanism, not just description)

3. **Hume's problem dissolves** (induction IS pattern recognition, which is basic and self-justifying—no regress, no need for external justification)

4. **Pattern recognition recovers all deductive functions** (truth-preservation, consistency, consequences, justification, rigor, knowledge-transmission, conceptual clarity—deduction is special case)

5. **Formalization is secondary** (Euclid/Chomsky codified patterns humans already recognized—useful but not necessary for rational capacity)

6. **Linguistic competence = UG + X** (UG provides recursive syntax via pattern recognition; X provides non-recursive semantics; both necessary)

7. **Wittgenstein's paradox refuted** (rule-following = pattern instantiation, which is direct and non-regressive; normativity from patterns themselves)

The Fundamental Insight

The tradition mistook a special case (recursive/deductive reasoning) for the whole of rationality.

Once we recognize that rationality is fundamentally pattern recognition, numerous pseudo-problems dissolve, and we can understand:

- How discovery actually works

- How knowledge is actually justified

- How language actually functions

- How minds actually think

- How AI actually succeeds

The Arc of the Entire Book

Part I: Proved that the class of recursive logics is not recursively enumerable

Part II: Showed this means formalism cannot formalize its own boundaries

Part III: Extended to language: natural languages are hierarchies with non-recursive semantic operations

Part IV: Identified the false equation (Rational = Recursive) and showed its catastrophic consequences across multiple domains

Part V: Developed the positive account: Rationality is pattern recognition, which is broader, richer, and more powerful than the deductive paradigm allowed

The Revolutionary Implication

Rationality cannot recursively characterize its own scope.

Just as:

- Formalism cannot formalize its own boundaries

- Recursion cannot recursively enumerate itself

- Language cannot linguistically characterize language

So too:

- Rationality transcends recursive specification

This is not a limitation but a liberation. We can stop forcing everything into recursive molds and recognize the full breadth of rational human capacities—including the non-recursive ones that make science, language, thought, and intelligence actually work.

Rationality is broader, richer, and more powerful than the deductive paradigm allowed. It's time to break free from the stranglehold of recursion.

End of Part V (Conclusion)

The book's central message:

A mathematical theorem (the class of recursive logics is not r.e.) reveals a deep structural feature of rationality: it cannot recursively characterize its own scope. This dissolves centuries of pseudo-problems, vindicates modern AI, liberates multiple philosophical domains, and establishes pattern recognition as the foundation of rational thought.

Rationality is seeing, not calculating.

That's what we've proven. That's what changes everything.